"What a hel[...]
important is[...]
many people[...]
are to the p[...]
emerge from[...] also has
an appealing practical side to help readers connect the
material with issues they may be dealing with personally,
as well as with what they are reading in the newspaper. I
highly recommend it!"

—SCOTT B. RAE, PH.D.
Professor of Christian Ethics
Talbot School of Theology, Biola University

"This booklet . . . does a wonderful job of explaining the
science behind genetic, stem cell, and cloning research, as
well as considering the theological implications of such
research. I recommend this resource to those who want to
know more about the facts of biotech research, to Christians who want guidance in interpreting what these scientific endeavors mean for their faith, and to anyone who
wants to gain a deeper understanding of the Christian view-
point on modern biotechnology."

—DAVID A. PRENTICE, PH.D.
Professor, Life Sciences, Indiana State University
Adjunct Professor, Medical and Molecular Genetics,
Indiana University School of Medicine

Basic Questions on
Genetics,
Stem Cell Research,
and Cloning
Are These Technologies Okay to Use?

The BioBasics Series provides insightful and practical answers to many of today's pressing bioethical questions. Advances in medical technology have resulted in longer and healthier lives, but they have also produced interventions and procedures that call for serious ethical evaluation. What we can do is not necessarily what we should do. This series is designed to instill in each reader an uncompromising respect for human life that will serve as a compass through a maze of challenging questions.

This series is a project of The Center for Bioethics and Human Dignity, an international organization located just north of Chicago, Illinois, in the United States of America. The Center endeavors to bring Christian perspectives to bear on today's many difficult bioethical challenges. It develops book, audio, and video series; presents numerous conferences in different parts of the world; and offers a variety of other printed and computer-based resources. Through its membership program, the Center provides worldwide resources on bioethical matters. Members receive the Center's international journal, *Ethics and Medicine,* the Center's newsletter, *Dignity,* the CBHD Internet News Service (weekly and monthly), special notification of Center conferences and publications, discounts on all Center conferences and resources, and more.

For more information on membership in the Center or its various resources, including present or future books in the BioBasics Series, contact the Center at:

The Center for Bioethics and Human Dignity
2065 Half Day Road
Bannockburn, IL 60015 USA
Phone: (847) 317-8180 Fax: (847) 317-8101
E-mail: info@cbhd.org

Information and ordering is also available through the Center's World Wide Web site on the Internet: http://www.cbhd.org.

BioBasics Series

Basic Questions on
Genetics, Stem Cell Research, and Cloning

Are These Technologies Okay to Use?

Linda K. Bevington, M.A.
Ray G. Bohlin, Ph.D.
Gary P. Stewart, D.Min.
John F. Kilner, Ph.D.
C. Christopher Hook, M.D.

Kregel
Publications

Basic Questions on Genetics, Stem Cell Research, and Cloning: Are These Technologies Okay to Use?

© 2004 by The Center for Bioethics and Human Dignity

Published by Kregel Publications, a division of Kregel, Inc., P.O. Box 2607, Grand Rapids, MI 49501.

ISBN 0-8254-3075-5

Printed in the United States of America

1 2 3 4 5 / 08 07 06 05 04

Table of Contents

Stem Cell Research and Cloning

Contributors

Linda K. Bevington, M.A., is Director of Research at The Center for Bioethics and Human Dignity, Bannockburn, Illinois.

Ray G. Bohlin, Ph.D., is President of Probe Ministries, Richardson, Texas.

C. Christopher Hook, M.D., is Director of Biotech Ethics at The Center for Bioethics and Human Dignity, Bannockburn, Illinois; and Director of Ethics at the Mayo Graduate School of Medicine, and Assistant Professor of Medicine at the Mayo Medical School, Rochester, Minnesota.

John F. Kilner, Ph.D., is President of The Center for Bioethics and Human Dignity, Bannockburn, Illinois.

Gary P. Stewart, D.Min., is a military chaplain serving with the U.S. Marine Corps.

Introduction

On April 14, 2003, U.S. scientists working in collaboration with an international team of researchers announced that they had reached their goal of sequencing the *human genome*—the genetic blueprint for a human being.[1] This "genome," or "blueprint," is comprised of small but intricately complex molecules of human DNA, which determine the individual hereditary characteristics of human beings. The deciphering of the human genome has been portrayed as an achievement greater than splitting the atom or going to the moon.[2]

As researchers gain an increasing ability to study, marvel at, and manipulate the human genome, what will they learn about human beings? How will they use the information that they glean? Who will benefit most from their research? Will our increased knowledge of genetics further our ability to create *transgenic* organisms—those containing genes from more than one species—for the purpose of medical research? Will it allow us to solve the organ shortage crisis by transplanting animal organs into humans (a process known as *xenotransplantation*)? The information that we stand to gain from genetic research is indeed staggering, and the way in which this knowledge is handled will depend upon the values of those who wish to apply the research findings and the convictions of those who develop policy pertaining to the acquisition and use of genetic data.

Genetic research and its impact on the future of humanity may indeed prove to be awe-inspiring. Yet other emerging technologies may carry an equal—or even greater—capacity for affecting human beings. In November of 1998, two teams of U.S. scientists announced that they had succeeded in isolating and culturing *human embryonic stem cells* in the laboratory.[3] This previously elusive feat, followed by unprecedented advances in "adult" stem cell research, have prompted many people to believe that stem cells hold the key to treating or curing some of the most devastating human afflictions (such as Alzheimer's disease, Parkinson's disease, diabetes, and heart disease).[4] Not long after the November announcement, some scientists began increasingly to vocalize their interest in producing cloned human embryos—and then sacrificing them for their stem cells— as the best means of improving the human condition. Though no medical therapies have yet been obtained, a team of South Korean scientists announced in February 2004 that they had cloned a human embryo and extracted its stem cells.[5] Other scientists have expressed enthusiasm for creating and bringing to term human clones so as to provide grieving or infertile couples with a child. Several claims have even been made that the birth of the first human clone is imminent— or has already occurred.[6]

If such biotechnologies indeed have the potential to achieve great good, should we not wholeheartedly encourage their progress? Is there a chance that great harm, rather than the intended benefits, might come to human beings as a result of such interventions? While the therapeutic potential of genetic technology, stem cell research, and cloning may prove to be great, these technologies do also present us with the possibility of great peril. If the innate dignity of all human beings is not protected, the endeavor to better human life will come at a cost to certain individuals (likely those who are weak or vulnerable). If, however, each human being is regarded as having immeasurable worth simply because he or she is human, scientists may seek to develop new medical therapies and alleviate suffering in a manner that upholds human dignity. We applaud the desire to help people achieve and maintain good health; however, such a pursuit must be achieved only by means that are ethical.

As Christians journey through the present and future biotechnological age, they must stand firm on the fact that human beings are not merely physical—but are also spiritual—beings. While emerging technologies may enable people to enjoy a healthier life on earth, everyone must eventually face physical death and the realities of existence beyond the grave. We should therefore look primarily to Christ for the ultimate repair of the human condition (John 10:10; 1 Cor. 11:23–26; 2 Cor. 5:21), while also encouraging ethical research in keeping with His mission of compassionate healing that He modeled during His earthly ministry.

This booklet addresses the hopes and concerns raised by genetic technology, stem cell research, and cloning. It provides information that people who live in technologically advanced societies should understand and respond to appropriately if the dignity of all human beings is to be protected equally. Christians, especially, should seek knowledge about these matters so that they can share with others God's truth and its implications for these challenges.

This book is not intended to reproduce all the available information on the subject but rather to simplify, complement, and supplement other available sources that the reader is encouraged to consult. Some of these materials have been listed at the end of this book. This book is not intended to take the place of theological, legal, medical, or psychological counsel or treatment. If assistance in any of these areas is needed, please seek the services of a certified professional. The views expressed in this work are solely those of the authors and do not represent or reflect the position or endorsement of any governmental agency or department, military or otherwise.

1. What is a gene?

Genes are the basic units of heredity. They consist of stretches of *DNA* (deoxyribonucleic acid) that determine certain characteristics such as hair color, eye color, and skin color. Combinations of genes enable the performance and regulation of all human functions. According to some of the most prominent researchers in the field of genetics, human beings have approximately thirty thousand to forty thousand genes located on the twenty-three pairs of *chromosomes* typically found in the nucleus of all but the reproductive cells (eggs and sperm) of our bodies.[7] (Reproductive cells have half the number of chromosomes, one from each pair.)

The DNA molecule is a double helical structure resembling a spiral staircase composed of sugars, phosphates, and two strings of *nucleotides* that run in opposite directions. DNA contains four nucleotides—abbreviated A (for adenine), G (for guanine), C (for cytosine), and T (for thymine). These two strands of DNA are complementary in that As should always pair with Ts from the opposite strand, and Cs should always pair with Gs from the opposite strand. The human genome consists of three billion of these *base pairs*. Because of the normal complementary pairing of the nucleotides, the sequence of one strand should automatically determine the sequence of the other strand.

In the 1960s, discovery of the genetic code revealed that a gene is comprised of a series of what are called *codons*. A codon is made up of three nucleotides in sequence. Codons specify the production of particular *amino acids*,

15

which are the building blocks of proteins. Our cells contain complicated and sophisticated molecular machinery that allows for the manufacture of specific proteins, which correspond to certain amino acid sequences. The sequence of amino acids in a protein is the primary determinant of the protein's function (for some examples, see the answer to question 10).

Think of the nucleotides of DNA as being like the letters of the alphabet. Just as a combination of letters form a word, so a combination of three nucleotides forms a codon (which codes for a specific amino acid). If a person strings words together in a particular order, the words form a sentence. The same is true of nucleotides—if a group of codons are strung together in a particular sequence, the codons form a gene (which in turn codes for a protein or protein component). Genes, then, are packets of information comprised of sequences of DNA that, through their protein products, code for specific biological functions. The more we learn about genes, the more we'll come to understand the functions of proteins in cells, allowing us a better understanding of the complexity of human life.

In 1953, James Watson and Francis Crick discovered the structure of the DNA molecule, revealing it to be much simpler than many had imagined. DNA's relative simplicity, though, makes its magnificence all the more evident. An extraordinary amount of genetic information vital to human functioning is packed into a molecule that is unbelievably small. To envision just how thin the strands of DNA are, imagine that the metal contained in the head of a pin has been drawn into a wire so thin that it can circle the equator thirty-three times![8] To grasp how much genetic information is packed into these amazingly thin filaments, think of the amount of information present in one thousand fine-print books of five hundred pages each. This is how much information is stored in the DNA located in a single human cell![9] Perhaps even more astounding, five or six thousand million times that amount of information is stored in the DNA of every human being on earth. Yet, if compressed, a person's DNA could be housed in just two aspirin tablets![10] Francis Collins, Director of the National Human Genome Research Institute, has stated that studying the DNA molecule "gives us a glimpse into the elegant way God thinks!"[11]

2. What is the Human Genome Project (HGP)?

A person's *genome* consists of all of his or her genes, the vast majority of which are organized into *chromosomes* in each cell's nucleus (see answer to question 1). The goal of the Human Genome Project (HGP) is to determine the location and nucleotide sequence of the estimated thirty thousand to forty thousand human genes housed on the twenty-two autosomal chromosomes and the two sex chromosomes. (Non-nuclear DNA is found in the cellular structures known as mitochondria and is also being studied by scientists working on the HGP.)[12]

The Human Genome Project was a multinational research project that, as a first step, determined the location of *genetic markers* on the chromosomes. A genetic marker is a particular sequence of DNA, which is found in the same location on the same chromosome in almost every individual. Genes are located on chromosomes by first approximating their nearness to known markers. Research on markers was completed in 1993, with the location of more than ten thousand markers. Second, the project produced a physical map of purified overlapping sequences of human DNA. This step was largely completed by 1997. The final step was to determine the entire sequence of all three billion base pairs and, ultimately, the sequences of all human genes.[13]

On June 26, 2000, U.S. President Bill Clinton announced that the National Human Genome Research Institute, along with Celera Genomics Corporation, had completed an initial sequencing of the human genome—an astounding feat that was accomplished two years ahead of schedule and under budget.[14] This draft sequence, consisting of approximately a 90 percent complete sequence of the human genome's three billion base pairs, was published in the journals *Science*[15] and *Nature*[16] in February 2001. On April 14, 2003, the International Human Genome Sequencing Consortium announced that the Human Genome Project had been successfully completed (also more than two years earlier and at less expense than expected), making available a finished sequence of the entire human genome.[17] According to National Human Genome Research Institute Director Francis Collins, "The true payoff from the HGP will be the ability to better diagnose [and] treat and prevent disease . . . [as we become] empowered to pursue those goals in ways undreamed of a few years ago."[18]

3. How might I be affected by the "genetic revolution"?

It has been asserted that the impact of the genetic revolution will surpass the effects of the industrial, atomic, and computer revolutions combined.[19] This amazing claim stems from our expanding knowledge of genetics, along with scientists' increasing ability to change the characteristics of living organisms (including human beings) by modifying their genetic makeup. As we acquire a more complete knowledge of human genes and the proteins they create, we will gain a greater understanding of the extent to which our individual genetic codes matter in determining particular traits and behaviors. As this knowledge is coupled with the ability to change the genetic code, many people will likely desire—and perhaps even feel compelled—to alter or determine their and/or their children's genetic sequences. Parents may even feel that it's their responsibility to do whatever they can to ensure that their offspring do not carry harmful genes, and may even take steps to bestow upon their children "designer" genes believed to give them certain advantages in life (see answers to questions 21 and 22). The possibilities for changing what we don't like about our children and ourselves are predicted to expand dramatically.

The genetic revolution will also alter our understanding of and response to human disease. As scientists learn more about the genetic basis of various illnesses, a whole host of challenging questions will be raised. If, for example, two people are known to be "carriers" of a gene that causes a devastating disease, should they get married—even though their future children would be at known risk for having the disease? If they marry, should they try to avoid having children? If they do become pregnant, should the child be genetically tested in the womb to determine if he or she has inherited the disease (see answer to question 18)? If tests reveal that the disease is present, should the child be allowed to develop to term, or should he or she be aborted? If a couple chooses not to terminate the lives of children determined to have the disease, should society be expected to help pay for the children's expensive lifelong care? Should health insurance companies be expected to provide coverage for such children, even though they will almost certainly require costly medical attention (see answer to question 14)? Should expensive gene therapy (technology being developed to alleviate the effects of, or even cure, genetic disease—see answers to questions 16 and 17) be available to such children? Should insurance companies or the government be responsible for paying for this potentially life-saving technology—even if an affected child could have been aborted? Could academic scholarships and vocational positions be legitimately withheld from such children in the future as a result of their poor health prognosis? What role should the government play in influencing these decisions?

The above paragraphs only begin to illustrate that we all—at one time or another—will likely be affected by the increasing power of genetic knowledge and technology. Whether we ourselves must personally make choices about the use of such technology or we are asked to vote for candidates who will pass genetic legislation, we need to be prepared to make informed decisions. Given the enormity and personal nature of the unfolding genetic revolution, we can't afford to be ignorant of the issues that will almost surely confront us.

4. Do our genes determine who we are?

The question about the role our genes play in shaping who we are takes us back to the old *nature versus nurture* argument. For decades, academics have argued over whether our genes or our environment is most influential in determining who we are as individuals. The pendulum of opinion has swung back and forth repeatedly.

It appears that *nature* may currently have the upper hand. The nearly constant influx of new data increasingly suggests that genes play an important role in our lives. Presently it seems that some sort of link may exist between particular genes and certain personality traits, intelligence, and tendencies toward violently aggressive behavior. Crude estimates indicate that general intellectual ability and personality traits such as extraversion/introversion, neuroticism, and conscientiousness may have as much as a 50 percent genetic component. While this is a far cry from saying that a single gene exists for intelligence or shyness, a genetic influence nevertheless appears likely.[20] Recently, the genetic makeup of individuals has been cited in attempts to excuse murder, rape, arson, exhibitionism, and the making of obscene phone calls.[21]

Given the current focus on the role genes play in shaping an individual, it is important to question the legitimacy of some of the claims being made about the powerful contribution of our genes. In their book *The DNA Mystique: The Gene as a Cultural Icon,* Dorothy Nelkin and M. Susan Lindee suggest that our society has regarded the influence of our genes as being more weighty than that which has been demonstrated scientifically.[22] Affirming this assertion, Francis Collins, Director of the National Human Genome Research Institute, has stated that the sweeping notions of genetic determinism are not consistent with scientific reality.[23]

Moreover, studies on identical twins that reportedly demonstrate a link between genetic makeup and homosexuality have, ironically, served to demonstrate the *limitation* of genetic influence on homosexuality (see answer to question 5). If a definitive correlation existed between a gene or genes and homosexuality, *all* pairs of identical twins (who have 100 percent of their genes in common) would be expected to have the same sexual orientation—which is not what the studies show.

Even the most fervent genetic determinist will acknowledge that a person's family, neighborhood, culture, and other life experiences can mold and shape his or her personality in a very significant way. Furthermore, it has been demonstrated that the expression of many genes is itself also substantially influenced by the environment. Consensus is lacking, though, as to how the two forces of genetics and environment work together to shape an individual's unique traits and behaviors. In addition to these influences, it is important to recognize that the spiritual component of a person may also temper his or her thoughts and actions.

In our genetics-crazed culture, we must exercise caution before attributing the presence of a certain characteristic (or even disease) to a particular gene or genes since such an explanation is most often overly simplistic.

5. If a behavior or trait is genetically based, does that make it morally acceptable—or at least excusable?

It is one thing to identify the genes involved in known genetic diseases such as cystic fibrosis, Huntington's chorea, or breast cancer, but what about the identification of genes that may play a role in determining intellect or social behavior? Wouldn't such a discovery prompt parents and teachers to raise or lower their academic expectations of certain children, based upon knowledge of their genetic makeup? Might it potentially eliminate, in some cases, personal responsibility for behaviors currently viewed as antisocial or even criminal?

We are witnessing a growing tendency to use genetic information as a rationale for avoiding personal responsibility. We need to take a step back, though, and ponder the legitimacy of such a tactic. It's true that genes, and their reported effects on various human characteristics, are being identified at a rapid rate. Nevertheless, we are still *far* from understanding how and to what extent most genes actually influence human traits and behaviors. Consider for a moment that scientists still have little to no idea how different genes work together to produce the human nose, or the shape of a person's eyes, or even the color of his or her skin—let alone a behavior such as violent aggression. While researchers certainly understand many of the biological processes involved, a complete connection between a person's genetic structure and his or her various traits and behaviors is, in most cases, nowhere close to being established.

Scientists have demonstrated that if they remove (or "knock out") a certain gene in an organism, a particular biological function will cease entirely. This does *not* mean, however, that this single gene is *solely* responsible for that function. On the contrary, a great number of other genes—not to mention non-genetic influences (e.g., nutrition, presence of carcinogens, etc.)—may also be essential for carrying out this particular function. Such experiments prove only that a given gene is certainly necessary for a specific function.

The interaction between multiple factors will likely prove to be even more evident in the case of behaviors. For a particular behavioral expression, for example, a complicated set of dozens or hundreds of genes may be involved, in addition to environmental factors (e.g., parenting, education, etc.). Cases in which a single gene can be said to determine a behavior are probably nonexistent. So even in a purely materialistic sense—without taking the role of human will and choice into account—the notion of strict "genetic determinism" is a myth (although we acknowledge that there may be exceptional situations in which genes have an unusually strong impact). Nevertheless, claims abound—and are often propagated by scientists and the media—that a person's behavior or other non-physical trait is due to his or her genetic makeup.[24]

In 1993, molecular geneticist Dean Hamer announced that he had discovered a region of the X chromosome (one of the two sex chromosomes) that is associated with homosexuality.[25] Earlier studies examining the potential link between brain structure and the incidence of homosexuality in identical and fraternal twins had raised the possibility of a link between genes and sexual orientation.[26] These studies were inconclusive, though, since they couldn't determine if environmental factors had mainly influenced the behavior of twins raised together, if behavior was indeed shaped by brain structure (and ultimately genetics), or if brain structure was itself affected by behavior. Hamer, however, boldly stated that he was "99.5 percent certain that homosexuality is genetic" based on the results of his research (though comments to his scientific peers were more nuanced).[27] Despite Hamer's expressed confidence, a follow-up letter in the journal *Science* criticized his conclusions as relying heavily on assumptions, as well as on a questionable use of statistics.[28] Moreover, a subsequent study reported to have found no such link between the X chromosome and homosexuality.[29] These claims and counterclaims clearly illustrate the lack of consensus regarding the interrelationship of genes and sexual behavior.

Setting the above studies aside, suppose that a gene allegedly linked to male homosexuality were found on the X chromosome. This would not mean that *all* men with the gene would be homosexual. We know that even identical twins are not uniformly gay or uniformly straight—and they share all of the same genes! Similarly, all men whose hypothalamus (region of the brain) contains a particular structure of cells reportedly linked with homosexuality are not gay. This tells us that non-genetic factors *must* be involved in determining sexual orientation. A genetic predisposition toward homosexuality does not, therefore, absolutely dictate that a person will be homosexual.

Even in cases where a genetic basis for a trait or behavior is conclusively demonstrated, it's doubtful that society will turn its back and simply overlook or excuse certain emotional traits or behaviors deemed to be socially unacceptable. Christians, especially, will likely not do so because they recognize that human fallenness extends to all aspects of human beings, including their genetic make-up. The acceptability of a certain behavior hinges not on whether it has a genetic basis, but on whether it conforms to God's standards. If any physical, genetic, or other limitation predisposes a person to do something that is contrary to how God intends for people to live, then there is a moral mandate, to whatever degree it is ethical and within human control, to change or overcome that limitation rather than invoke it as a justification for unacceptable behavior. If someone has a bent toward hurting people—whether environmentally, genetically, or otherwise based—the infliction of injury does not become excusable just because it has a recognizable basis.

Those who are Christians should experience an increased capacity to overcome aspects of their fallenness by relying on the healing power of Christ, and they should encourage others to do the same. While people may understand that certain aspects of their behavior are due in part to their genetic makeup, such an understanding should not allow them to escape responsibility for their actions, provided (as will most often be the case) that they are able to control or influence them, or have ethical means available to assist them to do so.

6. What is genetic engineering?

The term *genetic engineering* often conjures up disturbing images. We might think of laboratory experiments resulting in the creation of Frankenstein-like monsters or the use of genetic information to create new social classes (such as those in the 1997 film *Gattaca*).[30] Genetic engineering can be generally defined, though, as

25

the manipulation or alteration of the genetic structure of a single cell or organism—usually with the goal of producing a desired effect.

Sometimes scientists will change the DNA sequence of a gene simply to study what effect the change has on the resulting protein product (see answers to questions 1 and 10). "Knock-out" experiments in mice, for example, seek to determine the effects of removing (or "knocking out") a particular gene from the mouse genome (see answer to question 5).

Recombinant DNA (rDNA) technology, on the other hand, typically involves taking a gene from one organism and "splicing" it into another organism of the same or different species, usually to achieve a certain benefit (see answers to questions 7 and 9). *Gene therapy* seeks to correct disease (see answers to questions 16 and 17) by introducing specific genes into a sperm or egg (*germ-line gene therapy*) or other cells (*somatic cell gene therapy*). Intervention at the genetic level may also seek to enhance physical or mental characteristics (see answers to questions 21 and 22). All of these alterations fall under the umbrella of genetic engineering.

It is important to recognize that genetic engineering is itself neither good nor evil. Rather, the nature and purpose of genetic engineering will ultimately determine whether a particular application is ethical.

7. How has genetic engineering been applied to bacteria and viruses?

Research laboratories around the world routinely use bacteria and viruses as the workhorses of nearly all genetic engineering research. Since the late 1960s, scientists have been taking genes from bacteria and placing them into other bacteria using *recombinant DNA (rDNA)* technology. In addition to their single chromosome, bacteria contain small circular pieces of DNA called *plasmids*. Easily removed from bacteria, plasmids can have new genes "spliced" into them and can then simply be reinserted into bacteria. Because this method readily allows scientists to determine the functions of particular genes, this one technique opened vast new areas of research. It was soon discovered that viruses also allowed scientists to isolate, reproduce, and study novel gene combinations. Viruses can be manipulated easily; thus, genes that are responsible for producing harmful or toxic proteins can be engineered out, thereby minimizing the risks involved in using them as research tools.

Bacteria and viruses continue to be used widely in genetic experiments around the world. Bacteria are also now being used as "production factories" for many pharmaceutical products. The first such product to be derived from bacteria was human insulin. Diabetics requiring insulin injections previously were forced to rely on insulin harvested from pigs slaughtered in meat-packing plants. Since the human body recognizes pig insulin as a foreign protein, patients often developed immune or allergic reactions, compromising the effectiveness of such treatment. In response to this problem, scientists used rDNA technology to produce bacteria that contained the gene for *human* insulin. As a result, the engineered bacteria produced human insulin as a waste product. Because the insulin was human in type, it was perfectly suitable for diabetic patients, thereby eliminating the need for pig insulin.

Human Growth Hormone and Factor VIII (needed for blood clotting in hemophiliacs) are also produced through genetically engineered bacteria. Human Growth Hormone (hGH) was previously only available from the pituitary glands of deceased humans and was therefore in short supply. Similarly, Factor VIII had been available only from human blood supplies—a less than ideal source, given the possibility of contamination. Another area of ongoing research is the engineering of bacteria designed to digest chemical and biological wastes, thereby greatly aiding waste management efforts.

8. Could genetic engineering result in the production of a dangerous "superbug"?

In the early years of genetic research using bacteria and viruses, scientists respected and were sensitive to the unknowns of genetic engineering technology. At various points, they self-imposed their own bans on such research and applied guidelines to different types of experiments, based upon the potential risks involved.[31] Such restrictions came to be regarded as unnecessary, though, since early experimentation never resulted in production of a dangerous "superorganism." As a result, many researchers became convinced that transferring DNA of known sequence and function into experimental organisms did not result in unforeseen harmful consequences, and the power and popularity of this technology soon grew. The present lack of concern on the part of some scientists regarding the application of genetic engineering to plants and animals may result from this initial positive experience with bacteria and viruses.

Such history does not, however, preclude the possibility that genetic engineering could be used with the intention of producing a "superbug." This technology might, in fact, be used to produce destructive agents of biological warfare that would be difficult to eradicate—a fear that may increase in our post–September 11 society. These agents might even be targeted to particular races and other groups who have identifiable genetic traits. Some have even speculated that HIV (human immunodeficiency virus), which causes AIDS, was intentionally produced through techniques of genetic engineering. Although this hypothesis has been successfully refuted, genetic engineering has nevertheless opened up new possibilities for evil.

Though this technology is indeed powerful, we cannot put the "genie" wholly back into the bottle. While genetic engineering could be used to carry out evil, that is not the fault of the technology itself, but of persons who would choose to use it that way. We must therefore do everything we can to ensure that genetic engineering is used to pursue good, instead of to harm or destroy human beings.

9. How has genetic engineering been applied to plants and animals?

Plants and animals have been genetically engineered in several significant ways. Genetic engineering in plants has focused on producing crops with certain valuable traits—such as improved food quality and resistance to disease, particular herbicides, or environmental stresses such as drought. Genetic engineering of animals has also pursued the production of better quality foods for human consumption. Most of the current animal research, however, is focused primarily on the use of sheep, pigs, and cattle as "biological factories" from which medical therapies for various human diseases are developed. The genetics of these species can be altered in a manner that causes the animals to produce chemicals and other products needed to treat human illnesses.

More specifically, some plants have been engineered to produce their own insect-repelling chemicals by inserting (via *recombinant DNA* technology—see answer to question 6) a gene from bacteria. Cotton, maize, and soybeans have all been modified this way. Crops have also been engineered to be resistant to certain weed-control chemicals, thus making it possible for farmers to use more effective and environmentally friendly herbicides without harming their crops. In addition, potatoes have been engineered with a gene from arctic fish that produces an "antifreeze" protein, making the potatoes resistant to frost. Moreover, tomatoes have been engineered in several ways to delay ripening, thereby increasing their shelf life. Research is also now underway to alter crops—such as potatoes, bananas, wheat, rice, and corn—that are important for the well-being of persons in undeveloped countries. Such crops are engineered to be disease-resistant or are fortified with various nutrients typically lacking in the diets of people living in these countries.[32] (So-called "Golden Rice," which produces vitamin A in the grain, is one such example.)

In October 2000, many U.S. citizens became more aware of the potential impact that genetically engineered foods might have on their lives. During that month, taco shells and other similar products made from genetically modified corn were recalled because, while the corn had been approved for animal consumption by the U.S. Department of Agriculture (USDA) and the Environmental Protection Agency (EPA), the Food and Drug Administration (FDA) had not yet approved the corn for human consumption. The recall was necessary to alleviate fears that the new protein produced by the engineered plants could result in dangerous and even deadly allergic reactions.[33] A March 2001 poll by the Pew Initiative on Food and Biotechnology indicated that 75 percent of Americans desired to know whether a food product contains genetically modified ingredients.[34] Similarly, a November 2001 Rutgers University Food Policy Institute study indicated that 90 percent of Americans believe that foods that have been subjected to genetic engineering should be specially labeled.[35]

Pigs and sheep have already been engineered to produce human proteins in their milk (see answer to question 29). These proteins are necessary for treating patients suffering from hemophilia, cystic fibrosis, cancer, diabetes, and many other diseases. The protein-containing milk is manufactured by inserting the human genes that are responsible for the production of these proteins into the animals' genomes. The proteins are then extracted from the animals' milk and used to benefit people with specific medical needs (see answers to questions 28 and 29).[36] Mice, rats, pigs, and sheep are also being engineered with specific human genes to serve as models in testing new therapies for human diseases. Pigs engineered in this way are being studied to determine whether they might be a source of organs suitable for transplantation into humans (see answers to questions 29 and 30).[37] Although many scientists believe that such research will lead to revolutionary advances in human health care, others have raised concerns about this type of experimentation (see answer to question 28).

Genetically engineering plants and animals is often thought to raise fewer ethical concerns than does genetically engineering human beings. Significant ethical issues are, however, raised by applying this technology to plants and animals. Regardless of the potential for improving human lives, we must take care to address such issues, instead of proceeding blindly down the path of scientific pursuit.

10. Why do people suffer from genetically caused illnesses?

"Why" questions are often two questions in one and can be very difficult to answer. The first of the two is the scientific question, wherein the "why" is best changed to "how." For our purposes the question then becomes, "How do genetic illnesses arise?" The second (and perhaps more challenging) question is moral in nature. By asking "Why do genetic illnesses occur?" people are often really asking, "Why does God allow people to suffer from such personal afflictions that occur through no fault of their own?"

The answer to the first question is simple: *mutations*. Mutations are mistakes in the DNA sequence. We all carry mutations in our DNA. Most mutations, though, have no effect either because they occur in genetic material that is not part of a gene or they do not significantly alter the corresponding amino acid and/or protein (see answer to question 1). Genetic disorders result when the function of a gene and its corresponding amino acid/protein is either altered or eliminated.

Sometimes a mutation is simply the *substitution* of one nucleotide for another. Nucleotide substitutions will change the codon, which will usually change the resultant amino acid sequence. (Some amino acids are coded by several codons. Changing from one codon to another may not change the amino acid sequence of the protein in these cases—see answer to question 1.) In sickle cell anemia, for example, the sixth codon in the beta hemoglobin gene, GAG, has been altered to GTG. The subsequent amino acid change causes a chemical modification of hemoglobin, the protein that carries oxygen in red blood cells. Red blood cells with this altered form of hemoglobin have a different, "sickle" shape. These sickled red blood cells will eventually become stuck in small capillaries, clogging them up. The clogged capillaries are the reason for this mutation's lethal effect.

Mutations can also result from a piece of DNA being deleted. A *deletion* may cause one or more codons to disappear. In the majority of people who have cystic fibrosis (CF), codon 508 out of 1,480 in the CF gene is missing, causing just one amino acid to be removed from the resulting protein. This one missing amino acid is sufficient to cause the severe respiratory and digestive problems of CF patients, which are often fatal before patients reach their thirtieth birthdays. Seventy percent of all CF patients have this mutation. The other 30 percent have a wide variety of different mutations in the same gene, all of which render the same effect.

Other disease-causing mutations arise when certain nucleotides are inserted into a genetic sequence; the *insertion* alters the way the genetic code in the remainder of the sequence is read. Nucleotide sequences may also be reversed, resulting in a mutation know as an *inversion*.

So far, genes associated with more than one thousand human disorders have been identified. They are located on all twenty-three chromosomes, as well as on the small circular piece of DNA found in mitochondria (an intracellular structure). Some scientists have estimated that as many as three thousand to four thousand human genetic disorders may be caused by defects in a single gene. Most disorders, however, appear to be due to the result of mutations in a host of genes. Several genes, for example, are known to be associated with Alzheimer's disease.[38] It's not known whether mutations in just one or several of these genes are necessary to cause the disease. Deafness may result from mutations in as many as 100 genes.[39] Thus, when it's reported that the gene for a certain cancer or other affliction has been discovered, what this typically means is that "someone has found *a* gene implicated in *some* forms of the disease, in *some* people, *some* of the time" (emphasis added).[40]

Other mutations result in the loss or gain of whole chromosomes. The loss of a chromosome in a sperm or an egg results in an embryo with only forty-five chromosomes (a condition known as *monosomy,* in which one of the twenty-three pairs of chromosomes is therefore incomplete).

Monosomy usually results in death before birth. The only chromosomal loss that embryos occasionally survive is the loss of the X chromosome in Turner Syndrome (90 percent of embryos with this condition die in the womb). Those who do survive are surprisingly normal, although they will be infertile. Another chromosomal mutation is a *trisomy*. Trisomy 21 (commonly known as Down's syndrome) is a disorder in which three (instead of the normal two) copies of chromosome 21 are present. Scientists still don't understand why the extra chromosome causes the severe mental retardation and other anatomical distinctions characteristic of this condition. Now that chromosome 21 has been sequenced, researchers hope that the information gleaned will result in fruitful research, helping them to understand and even cure Down's syndrome. Another type of chromosomal mutation is a *translocation,* in which a chromosome or fragment of a chromosome becomes stuck to another chromosome or chromosomal fragment, sometimes resulting in harmful consequences.

The moral question of why people suffer from genetically caused illnesses is different from the scientific question. It asks the question "Why?" in a more ultimate sense. Mutations—as well as other human tragedies—occur as a result of the fall of humanity into a sinful, and therefore mortal, condition. Scripture tells us that the serpent was cursed, Eve was cursed, and Adam was cursed (Gen. 3:14–19). *All* creation, however, according to Romans 8:18–22, was subjected to brokenness. It therefore groans and suffers, eagerly awaiting God's intervention to set it free from its slavery to corruption. Human disobedience produced a world that is not as God intended.

This does *not* mean, however, that a person with a genetic illness is being punished for a specific, individual sin. In many cases, a person's genetic mutation is present at conception before the opportunity to commit a sin is even present. Even in cases where a mutation arises later in life, we should not necessarily regard it as God's punishment of the one afflicted. Neither should we view genetic mutations in children as punishment of the parents. Asking why someone suffers from a genetic disease is ultimately no different than asking why someone was badly injured in a traffic accident from which others walked away unharmed.

Christians know that their suffering is temporary and that in the end God will bring all things together for their good (Rom. 8:28). The apostle Paul tells us that those who suffer are to share the comfort they receive from God with others who suffer (2 Cor. 1:4)—therefore, our suffering is also designed to benefit others (2 Cor. 1:6). Furthermore, we may suffer so that we'll learn to trust in God and not ourselves (2 Cor. 1:9). Suffering should also enable us to understand and appreciate more fully the immeasurable worth of human life (Eccl. 7). Even Paul suffered from some kind of frailty, which he repeatedly asked God to take away. God's answer to Paul was, "My grace is sufficient for you, for My power is perfected in weakness" (2 Cor. 12:9). Although we should not seek out suffering or regard suffering in itself as good, we should seek to discern (and be encouraged by) God's various purposes for it.

Although suffering often leads to positive outcomes, part of the Christian mission has always been to alleviate suffering whenever it's possible to do so by ethical means. Christ's miracles often provided relief from suffering and served as a reminder of what life will be like in His eternal kingdom. Furthermore, it's likely that He cried at Lazarus' tomb in part because He was moved by the suffering brought about by death and, ultimately, by sin. We should thus seek, to the extent that it is ethically possible, to apply

35

genetic technologies or other means to relieve or remove the suffering that results from genetic disease.

11. Should I be tested for a genetic abnormality?

There is no reason to undergo genetic testing unless it is suspected that you may have a genetic disorder or it's suspected or confirmed that a genetic disorder runs in your family. Genetic tests are expensive and are specific for a particular disease. Current technology does not allow us to test for a host of genetic conditions with a single blood test. A genetic test is not appropriate, therefore, unless good reasons exist to suspect that you may have a genetic disease.

If you are at risk for a genetic disorder and are considering testing, you should ask yourself an additional question: "Will the results of the test potentially alter how I live my life, in terms of daily routine and/or medical treatment?" If knowing that you have a specific genetic condition will not result in such changes (e.g., if no special diets are recommended or no treatments are available for a given disorder) and will not influence any reproductive decisions you make, then the test may be unnecessary.

In many cases, however, testing for certain genetic disorders can be quite beneficial. Hemochromatosis is the most common genetic disorder among Caucasians, as one in nine Caucasians are "carriers" (*heterozygous*, or having one copy of the disease gene) and one in three hundred actually have the disorder (*homozygous*, or having both copies of the disease gene). Hemochromatosis is a hereditary iron disorder that causes the body to retain excess iron in the blood. This surplus iron eventually gets stored in the liver and other organs, and symptoms typically begin appearing during adulthood. Some of the symptoms of iron overload include chronic fatigue, increased susceptibility to infection, liver function abnormalities, arthritis, diabetes, loss of sexual libido, impotence, infertility, swollen stomach, heart pain, shortness of breath, loss of weight, and decrease in body hair. Unless the excess iron is removed, the buildup can lead to cirrhosis of the liver and even liver cancer. Currently, the only treatment is to remove the surplus iron a little at a time through a phlebotomy (usually a pint of blood is drawn every one to two weeks) until the iron levels return to the low end of normal. This may take months or even years. Once the desired iron level is reached, decreasing the phlebotomy intervals to once every three to six months should be sufficient to control the iron levels in the blood. If a patient has not already suffered liver damage, he or she can be expected to live a normal life. Early screening and diagnosis can help patients avoid organ damage and premature death due to this disorder.

A genetic test for hemochromatosis is available, but for those who have already been diagnosed with the disease, the results will offer no new information. If, however, a person known to have hemochromatosis has children, it would be wise for them to be genetically tested for the disorder so that an early diagnosis can be made. If the test results are positive, the child's iron levels should be closely monitored so that less frequent phlebotomies can be instituted when appropriate, rather than the usual initial months of weekly or bi-weekly phlebotomies.

The decision to undergo testing should be yours and yours alone, with no coercion whatsoever from medical professionals or anyone else. You should also ensure (before undergoing a particular test) that you understand what you are being tested for and why. Finally, you should be permitted to determine whether or not the information obtained will be destroyed once it has been provided to you and whomever else you wish to have access to it.

While genetic testing should certainly be employed in some situations (such as in the previous example), decisions whether to be tested are often less than straightforward. In such cases, the possible advantages and disadvantages of undergoing testing should be carefully considered (see answers to questions 13–15).

12. What is genetic counseling?

Genetic counseling is a relatively new field that is becoming increasingly necessary. As our knowledge of genetics and our ability to intervene at the genetic level expands, more and more people will have genetically related questions about their health. Genetic counseling consists of a number of components designed to help persons address these types of concerns.

The first component is *diagnostic counseling*, which consists of gathering all the necessary information in an attempt to provide counselees with a definitive genetic diagnosis. This step involves obtaining the medical histories of counselees and their families and may involve administration of a genetic test, if one is available. While some disorders are well-characterized and understood, others are more complex and mysterious. A disease may have been inherited, thus raising the possibility of risk to future children and other existing family members, or it may be the result of a new mutation (see answer to question 10).[41]

The second component is *informative counseling*. During this phase, the genetic counselor seeks to help counselees comprehend the medical facts related to specific genetic conditions that they have been determined to have or be at risk for developing and/or that may afflict their future children. Such counseling is based on what is presently known about how certain disorders are transmitted and on information about a counselee's genetic makeup. The counselor also discusses both the immediate and future consequences of the genetic condition of concern. This component is perhaps the most crucial, as providing accurate information regarding prognosis is critical if counselees are to make good decisions about treatment and/or preventive measures. The genetic counselor also helps counselees to gather and understand as much information as possible about a particular disorder. Numerous resources are available on genetic diseases in general, as well as on disorders that typically affect certain groups of people. The genetic counselor can help direct counselees to those sources of information that are likely to be most accurate and helpful.[42]

Finally, the genetic counselor helps counselees to select treatment options that are consistent with the counselees' own values and beliefs. Although the field of genetic counseling generally requires counselors to support the decisions of their counselees—or (if this is not possible) to refer them to someone who can—Christians may find it especially helpful to seek a genetic counselor who shares or is at least sensitive to a Christian's moral foundation.[43]

The third component is *supportive counseling*. During this stage, the genetic counselor helps counselees deal with the emotional and physical consequences of learning that they or a family member have a particular genetic disorder. The genetic counselor strives to enable those who are personally afflicted with or affected by the disorder to make the best possible adjustments to the challenges they face. If, for example, a counselee will be or is already showing disease symptoms, the counselor offers information about the debilitating process of his or her disorder and provides him or her with practical suggestions for dealing with the affliction. This phase also entails ministering to counselees through the initial and subsequent stages of the grief process that may ensue upon learning a diagnosis, as well as pointing the way to support services such as intervention programs and genetic support groups.[44]

The final component is *follow-up counseling*. In this phase, genetic counselors continue to serve as a source of new information and emotional support when needed. While the number and frequency of sessions usually diminishes over time, genetic counselors stand ready to assist their counselees should new questions or challenges arise.[45]

Genetic counseling can be a tremendous field for Christians to enter. Because of the many advances in genetic technology, the demand for genetic counselors and their potential to assist people in need will likely expand rapidly. There may be no more strategic place in the coming years to administer grace and truth to those suffering from illness than in the office of the genetic counselor.

13. How can genetic information be used to help me?

The information generated by the Human Genome Project (HGP) (see question 2) is predicted to comprise the sourcebook for biomedical science in the twenty-first century and to be of immense benefit to the field of medicine. It should help us to understand and eventually treat many of the more than 5,000 genetic diseases that afflict humankind, as well as the many multifactorial diseases in which genetics play an important role. As the amount, quality, and specificity of genetic information continues to increase, the level of control over personal health should also become greater.

The HGP's accomplishments are already very promising. Several genetic illnesses are now treatable through gene therapy. Genes have been isolated for a number of devastating conditions, including amyotrophic lateral sclerosis (Lou Gehrig's disease), cystic fibrosis, Duchenne's muscular dystrophy, Fragile X syndrome, Huntington's disease, neurofibromatosis, retinoblastoma, retinitis pigmentosa, and Wilms' tumor. Discoveries of genes for other diseases are expected to follow in great number. Once the disease genes are identified, efforts can be made to develop treatments or even cures for these disorders.

The results of genetic testing can be of tremendous value to the individuals being tested, as well as to their families. As illustrated in the answer to question 10, children at known risk for hemochromatosis can have their minds put at ease if their genetic test results are negative. If the results are positive, proper monitoring of iron levels can eliminate the need for physically exhausting future interventions and also help prevent the development of iron-induced liver disease and other symptoms. In a similar vein, advance knowledge of a genetic risk for colon cancer (for example) can lead to preventive dietary measures and regular checkups to enable detection and removal of precancerous polyps. Such practices may result in a life free of colon cancer in later years. Furthermore, if a person suspects that he or she may carry a gene for a serious or fatal disorder, genetic testing may better equip him or her to make decisions regarding marriage and family (see answer to question 3).

Genetic information may indeed empower people to gain better and more effective control over their medical futures—provided that it is appropriate to obtain such information (see answer to question 11) and that people understand enough to respond to it constructively (see answer to question 12). In order to ensure that its promise for good is not undermined, genetic information should always be employed in a manner that upholds the innate dignity of human life.

14. How can genetic information be used to hurt me?

The results of genetic testing may, indeed, provide people with new and powerful information, allowing them to have greater control over their health. Alongside these benefits, though, genetic information can become an instrument of discrimination and prejudice and a source of personal fear and anxiety.[46]

Genetic information has already been used to prevent people from obtaining both health and life insurance and from finding employment. In the 1970s, many African Americans who were known to have the sickle-cell trait lost their jobs and/or their health insurance—although they didn't actually have sickle-cell anemia (see answer to question 10) and were otherwise healthy and at no risk of developing complications! In another case, a patient with hemochromatosis was denied health insurance, even though this disease can be well managed without the need for costly medical treatments (see answers to questions 11 and 13). Genetic discrimination was also evident in a case in which an HMO indicated its unwillingness to pay for the delivery and health care of a fetus who was known to have cystic fibrosis, although no qualms were voiced about paying for an abortion.[47]

All too often, discrimination occurs when an employer or health/life insurance provider possesses an incomplete understanding of genetic information. It's important to recognize the difference between actual disabilities and potential ones. When applying for jobs that require good health, people generally shouldn't be denied a position based merely upon genetic (or any other) information that indicates a health problem or disability *may* develop. Only if and when such a development actually occurs should their suitability for employment be reevaluated.

Genetic information may also lead to individual and family stress. If, for instance, a young adult receives information that he or she has the genetic mutation that causes Huntington's disease (a serious disorder of the central nervous system characterized by physical and psychological deterioration usually evidenced in a person's forties or fifties), the resulting emotional stress can lead to depression as well as to dramatically negative changes in lifestyle. In one survey of persons with reasons to be tested for Huntington's, 11 percent said they would consider suicide if they tested positive and 5 percent said they would certainly commit suicide. Forty percent of people testing positive for Huntington's develop depression, while others turn to sexual promiscuity and impulsive behaviors. Many at risk for Huntington's disease understandably choose not to be tested.[48] The possibility of negative repercussions of certain test results calls for both wisdom and caution. Such is especially true in the case of children who are unable to comprehend the potential benefits and risks of obtaining such information.

As genetic knowledge becomes more available, future marital decisions may increasingly require some level of genetic disclosure. Surely someone who is at known risk for, or has been diagnosed as having, the Huntington's mutation should disclose this to a potential spouse. Even the disclosure of something as manageable as hemochromatosis would show love and respect for a prospective spouse by allowing him or her to consider more fully the potential health consequences for children that their union might produce. (If both potential marriage partners are "carriers," each of their children will have a 25 percent chance of receiving both genes and therefore having the disease. If one or both of the couple actually have the disease, the risk to future children would be 50 percent and 100 percent, respectively.) Some people may not want to risk losing a potential spouse by revealing genetic information; however, it's better and more loving to be honest than to risk causing someone the pain of betrayal and broken trust should a genetic illness later be manifested in his or her spouse or in the couple's children.

Sometimes informing one's family that a genetic test has been done, or has returned positive for a particular condition, will produce negative reactions. Close relatives may initially be more concerned about the inevitable consequences such news might bring for their *own* lives than they are about the suffering of their loved one. They may be especially fearful that they, too, will develop the disease and/ or are at risk for passing on a potentially harmful gene to their future offspring. Even if such fears are not present, the initial shock of hearing "bad news" about the welfare of a family member can be overwhelming. Although the patient will suffer physically, spiritually, and emotionally from his or her disorder, the family who must care for the patient will also endure significant suffering. In a real sense, the disease of one family member is the disease of the whole family. For this reason, some would prefer to remain ignorant of even the potential for genetic disease in a relative and would, therefore, resent being informed. It is our hope, though, that an increase in general knowledge about genetics, as well as about methods of caring for those with genetic disorders, will occur in the population as a whole—thereby alleviating some of these negative family situations. Genetic counseling (see answer to question 12) can be tremendously helpful toward this end.

It is likely that federal policy will become necessary to increase the likelihood that genetic information will be used to benefit, and not harm, people. Life is sacred—both yours and your neighbors (Matt. 22:37–40). Decisions regarding genetic information should therefore, to the extent possible, take into consideration the welfare of all.

15. Who should have access to my genetic test results?

This will certainly be an increasingly contentious issue in the coming years. Currently, there is not a great deal of genetic information available since only certain individuals have been tested and only for specific diseases. Physicians and health care professionals will, however, likely rely more frequently on genetic information to care for their patients. There is already an underlying fear regarding public access to present—and especially future—stores of genetic information.

Most people are frightened by the thought of the government having free access to data banks of genetic information gathered from the public. The eugenic abuses in the U.S. and Europe from 1920 through the 1940s are still fresh in many people's minds (see answer to question 23). The fact that health and life insurance companies have a legitimate interest in obtaining a person's genetic profile may also be alarming, and it is understandable why persons who are insured or are seeking to be insured may not wish to share such information.

It is reasonable, however, for health and life insurance companies to require people to provide information—genetic or otherwise—that will likely affect their future well-being. Insurance is about protecting against risk, and the greater the risk, the greater the cost of coverage; thus, companies need to gather information in an attempt to quantify risk.

Family members may also have a special interest in discovering the portion of a relative's genetic information that may carry implications for their own health, although some may prefer not to know such information—particularly if no treatment or cure is available for the disorder in question (see answer to question 14).

Some have taken active steps to ensure that their genetic information does not fall into the hands of others. While writing a story on genetic engineering for *National Geographic* magazine, author James Shreeve had his DNA tested for *SNPs* (*single nucleotide polymorphisms*), which are abnormal individual nucleotides in the DNA strand (see answer to question 1). SNPs can confirm a person's identity because each person's composite of SNPs is unique. A postage stamp-sized chip produced by Affymetrix was used in the testing process. This "GeneChip" could not search for the presence of actual genes, although it was anticipated that it could do so in the future. Lest his genetic information be accessed later, Shreeve arranged to have it deleted from the Affymetrix computers once he was finished viewing it.[49] While this strategy was effective for Shreeve, once such information is contained in a person's medical records, it will not be possible to discard it since many institutions indefinitely retain permanent files.

The details of your genetic makeup are indeed very personal. Safeguards should be in place to prevent a person's genetic (as well as other health-related) information from being used unjustly to deny or revoke employment, insurance, and/or health care.[50] The sharing of confidential genetic information should also be limited to protect patients from embarrassment, social stigmatization, and economic discrimination. We can be thankful that an earnest desire currently exists on the part of the medical community to limit the availability of such information.

16. What is gene therapy, and how is it carried out?

Gene therapy is a relatively new approach to treating, curing, or even preventing human disease by introducing specific genes into a patient's cells. Scientists have introduced genes into mice, cows, sheep, and pigs for years and have, more recently, begun applying this technology to human beings as well.

A common way of classifying gene therapy is to distinguish between *somatic cell* gene therapy and *germ-line* gene therapy. Somatic cell gene therapy targets a patient's non-reproductive cells, while germ-line gene therapy targets a patient's reproductive cells (sperm and eggs). Somatic cell gene therapy (see answer to question 17) changes only the recipient's genome, and not the genomes of future generations. Germ-line gene therapy, on the other hand, alters the recipient's egg or sperm cells with the intention that the genetic changes be passed on to his or her future children. Gene therapy that alters all cells of an embryo would by nature be germ-line manipulation because some of the treated embryonic cells will go on to produce the cells that will ultimately develop into the gametes (sperm and eggs). While both forms of gene therapy raise important ethical issues, germ-line gene therapy has yet to enter clinical trials because of its yet-unknown consequences for future generations.

Gene therapy may be used to alleviate the symptoms of a patient's genetic disease by transferring a non-defective copy of the defective gene into the tissues most affected by the genetic condition. Researchers are attempting, for example, to transfer a healthy version of the cystic fibrosis (CF) gene into the lungs of CF patients in order to coax their lungs to produce more normal mucus, thereby alleviating the debilitating effects of CF. This procedure, though, will not *cure* patients since their defective CF genes will remain, nor will it prevent patients from passing such genes to any children they might have. Gene therapy may one day be used to insert a healthy gene into a sperm or egg, or even into an early embryo known to have a particular genetic disease. Doing so, it is hoped, will prevent (or at least significantly reduce) the suffering that is typically experienced as a result of the disorder. We're still far, though, from understanding and controlling the effects of germ-line intervention; therefore, this type of genetic therapy should not now be attempted.

Various *vectors* including several types of viruses, DNA/lipid/protein combinations, and even artificial chromosomes[51] may be used to deliver a desired gene into a gene therapy patient. Scientists commonly use viruses in gene therapy protocols because of the unique ability of viruses to invade a cell and take up residence in the cell's DNA. In most clinical trials, some of a patient's blood or bone marrow cells are removed, cultured in the laboratory, and exposed to the gene-carrying virus. Once the cells have incorporated the gene into their DNA, they're again cultured in the laboratory and then injected back into the patient. This type of gene therapy is termed *ex vivo* (Latin for "outside the body") since the gene is inserted into the patient's cells while the cells are in a laboratory culture. Another method for carrying out gene therapy uses vectors to deliver a desired gene directly to cells in the patient's body. This type of gene therapy—which is less common than *ex vivo*—is called *in vivo* (Latin for "inside the body") gene therapy.[52]

While viruses may be particularly well-suited for use as vectors in gene therapy, they also may pose significant risks to the patient (as evidenced by the tragic death of Jesse Gelsinger—see answer to question 17). Because viruses typically have the capacity to infect more than one type of cell, a gene-carrying virus might alter cells other than those targeted. Moreover, the virus could cause a negative reaction in the patient (possibly even a deadly one, as in the Gelsinger case) or might be transmitted to other individuals or into the environment. While strict precautions are required as a means of avoiding such outcomes, these risks have prompted some scientists to look favorably upon the above-mentioned DNA/lipid/protein combinations and artificial chromosomes as possible alternatives to viral vectors.

17. Should gene therapy in human beings be pursued?

In 1991, the first gene therapy trial in humans was carried out in an attempt to correct a life-threatening immune disorder (severe combined immune deficiency, or SCID) in a four-year-old girl named Ashanti DiSilva. Doctors drew blood from Ashanti and treated her defective white blood cells with a gene responsible for producing a crucial enzyme that she lacked. The genetically modified cells were then injected back into Ashanti, with hopes that they would produce not only the enzyme but also future generations of healthy cells.[53] Although the girl is doing well today, the ultimate success of her gene therapy is still in question. While the treated cells did produce the needed enzyme, no healthy new cells were produced. Further, Ashanti has received subsequent rounds of gene therapy to maintain the level of enzyme in her blood, and she continues to take doses of the enzyme itself. Thus, it's unclear the extent to which her gene therapy is beneficial.

More recently, sixteen heart disease patients, who were likely close to death, received by injection straight into the heart a solution containing copies of a gene that triggers blood vessel growth. Due to the growth of new blood vessels around clogged arteries, all sixteen patients showed improvement and six were completely relieved of pain. In each of these cases, gene therapy was performed as a final option for treating a fatal condition. Full and informed consent was provided in each instance, either by the patients themselves or their parents.

Following these (and other) early successes that raised hopes for the promise of gene therapy, the entire field was dealt a severe setback in September 1999 with the death of eighteen-year-old Jesse Gelsinger. At the University of Pennsylvania, Gelsinger had undergone gene therapy for an inherited enzyme deficiency (ornithine transcarbamylase [OTC] deficiency).[54] The most challenging aspect of gene therapy has been determining the best method of delivering the healthy gene to the right cells so that the gene can be incorporated into the cells' chromosomes. A common method for carrying out such delivery is to use weakened forms of viruses, which naturally incorporate their genes into human cells (see answer to question 16). Jesse Gelsinger apparently suffered a severe immune reaction to the engineered virus used in his therapy and died four days after he was injected with it. This same virus had been used safely in countless other trials, but in this case, something went dreadfully wrong. Further investigation at the University of Pennsylvania's Institute for Human Gene Therapy uncovered procedural irregularities, prompting suspension of all ongoing trials at the Institute.[55] Other institutions were also found to have failed to disclose publicly the deaths of gene therapy patients, as is required in clinical trials. A Congressional review ultimately ensued, addressing various troubling aspects of gene therapy.[56]

Additional concerns about the safety of gene therapy were raised in 2003 when two other patients with SCID who had been treated in this way reportedly developed a leukemia-like condition, prompting the FDA to suspend twenty-seven gene therapy trials in the United States pending further investigation.[57]

These unfortunate mishaps indicate that the technical problems of gene therapy have not been fully resolved. Such a fact is an essential part of the information that must be conveyed to patients who are considering experimental gene therapy.

Other challenges to gene therapy include determining the function(s) of the estimated thirty thousand to forty thousand human genes, and understanding how they interact with each other and with the environment. High research costs and experimental regulations may also slow the progression of gene therapy research.[58]

To relieve suffering whenever it is possible to do so by ethical means is to follow in Christ's footsteps. Provided that gene therapy holds real promise for improving human health, we applaud and encourage its safe and ethical pursuit.

18. Should I consider amniocentesis or other forms of prenatal genetic diagnosis (PGD)?

Prenatal genetic diagnosis (PGD) represents a rapidly expanding service within obstetrical care. Physicians of an earlier era relied on the rate of uterine growth, the activity of the child, and a heartbeat detectable after five months to monitor the well-being of a fetus. Little else was discernable until birth. Today, physicians can perform *amniocentesis* (a procedure in which the developing child's amniotic fluid is obtained for testing) to detect numerous genetic conditions. Examination of the placental tissue via *chorionic villus sampling* (CVS) may also provide certain genetic and metabolic information. It is important to remember, however, that while some hereditary disorders can be diagnosed, few effective treatments are currently available.

In this era of high-tech assisted reproduction, sperm and

egg may meet *in vitro,* with the first cell division taking place outside the body. It is therefore now possible to test one of the cells of an early embryo to gain information about him or her prior to implantation. As scientists gain knowledge from the Human Genome Project (see answer to question 2), it's likely that technology will be able to predict from the earliest stages of life the chances that a person will suffer from particular medical conditions. The revelation that an embryo has a genetic disorder sometimes leads to an unfortunate recommendation of abortion, particularly when the diagnosis is obtained early in a woman's pregnancy.

In some cases, even the lives of healthy embryos may be terminated following genetic testing. If, for example, a couple has a critically ill child whose only chance for survival is a bone marrow transplant, they may seek to have another child whose bone marrow would (hopefully) be compatible with his or her ill sibling. Once this child is born and determined to be a suitable donor, some of his or her bone marrow cells could be extracted and transplanted into the ill sibling. In such scenarios, embryos whose bone marrow is demonstrated to be incompatible might be in danger of being discarded. Rather than being merely hypothetical or futuristic, such situations have, sadly, already become a reality—as evidenced in the U.S. case of the Nash family (2000),[59] as well as the U.K. case of the Whitakers (2001)[60] and Hashmis (2002).[61]

Before subjecting your unborn child to PGD, you should ask yourself several questions. First, you should identify and evaluate the importance of the benefits you expect to obtain from the resulting information. You should also inquire about the risks (e.g., of miscarriage) that would be posed to the embryo and seek to determine what level of risk is acceptable. While the risk of miscarriage is quite low (approximately one in five hundred to one in a thousand pregnancies with sonographically directed needle placement for amniocentesis, with a slightly higher risk [1–2 percent of all pregnancies] for CVS), it is nevertheless real. Parents should also consider whether and how the information gained from prenatal genetic testing would benefit *the embryo*, especially given that such testing poses risks to him or her.

19. How can we help those who suffer from genetic disorders?

Gene therapy (see answers to questions 16 and 17), as well as other forms of medical intervention, may indeed help improve the lives of those who suffer from genetic disorders. We should, then, encourage such people to seek out the best available medical and ethical counsel concerning their particular condition. We should also advise them to seek genetic counseling (see answer to question 12) and other forms of support when necessary.[62] In addition to helping people obtain medical and professional assistance, we should also personally seek to help meet the needs of those facing genetic challenges by caring for them both physically and emotionally, providing them with special housing or special food if needed, and/or simply offering a word of hope and life. Reaching out in each of these ways can greatly bless those struggling with a genetic disorder.

We must always be careful not to treat persons with genetic afflictions—no matter how severe they may be—as less than human. Because the lives of all human beings are sacred, we should do everything we can to ensure that genetically disadvantaged people are loved and accepted and that they receive adequate care. Christians who treat the "least of these" with dignity and compassion (Matt. 25:34–45) will go a long way in living out the church's mission to serve both God and people.

20. Should I be able to select the gender of my child?

One consideration in answering this question involves the timing of the selection itself. If a boy is desired and the gender is to be determined *after* fertilization, a "negative" result (meaning that a girl has been conceived) may prompt the parents to terminate their child's life either in the lab (if conception is achieved through *in vitro* fertilization) or in the mother's womb via abortion. Such gender selection should be prohibited for the following reasons.

Simply put, the life of an innocent human being is sacrificed. To terminate a life just because it is not the "product" one desires is demeaning and discriminatory. Each human being, male and female, is uniquely created, known, and valued by God (Job 31:15; Ps. 139:13–16; Isa. 49:1; Jer. 1:5; Gal. 1:15; Eph. 1:3–4). Furthermore, all human life exists primarily for God's pleasure and purposes, not ours (Col. 1:16). The sanctity of human life, created by God in God's image (Gen. 1:27; 9:6), demands that the life of the unborn not be ended for any reason except (some Christians would assert) possibly in rare cases when the sanctity of another human life (that of the mother) requires its termination. Even in a country where abortion is legal, it is hoped that restrictions would be put in place to prevent the taking of a life simply because that life is the "wrong" gender. It is not the fault of children that they are conceived as male or female, and nothing is inherently wrong with being either.

Another way to determine the gender of a child is to do so *before* fertilization. Procedures exist that reportedly separate sperm that carry the Y chromosome from those that carry the X chromosome.[63] Eggs fertilized by sperm carrying the Y chromosome will be male, and eggs fertilized by sperm carrying the X chromosome will be female (X plus Y produces a boy; X plus X produces a girl). If the sperm sample used to fertilize an egg in the lab has been selected for the Y chromosome, the odds are greatly increased (about a 90 percent likelihood) that a boy will result rather than a girl. (If a girl is born, it simply means that an error was made in the sperm selection process.) In keeping with the previous paragraph, anyone using this method must be willing to accept either a boy or girl and not discard the embryo or abort the baby if it is determined that he or she is not the preferred gender.

The basic question remains, however, as to whether use of such a method is a good idea in the first place. One proposed justification is to help ensure that a child will be free of a sex-linked genetic disease. Red or green color blindness, hemophilia A and B, and fragile X syndrome are all due to mutations on the X chromosome. Males are therefore much more likely to suffer from these conditions (in females, when one X chromosome carries the normal gene, the presence of the mutated gene on the other X chromosome often has no effect). If there is a known risk of one or more X-linked diseases in a family, opting for a girl by sperm selection can greatly reduce the possibility of having an afflicted child. The motive of reducing suffering is laudable, particularly when this can be achieved without forfeiting a life.

Gender determination by sperm selection is not, however, wise in all cases. Even when the purpose is to avoid a sex-linked disease, we run the risk of communicating to those who have uncorrected or uncorrectable genetic diseases that they would have been better off not being born. Although no disease is wanted, those who are afflicted by disease *are* wanted. Furthermore, while it may seem innocent enough for a couple who already have a son to wish to balance their family with a daughter, why is this so important? What fuels this desire? It is dangerous to think that we know more about life and God's purposes than God Himself. Moreover, the idea that we can control our destinies and that of our children is arrogant. There's a purpose for each child—boy or girl—that's far beyond our human capacity to grasp (Isa. 49:1; Jer. 1:5). Finally, genetically imposing characteristics upon children that are beneficial for all human beings is often regarded as controversial. Choosing for children a characteristic like maleness—which is *not* beneficial for all human beings to have—is the first step down a road that we should be hesitant to travel. It commits us to allowing parents or society to impose on a child whatever characteristics they wish, regardless of whether or not such attributes are good for him or her.

In some cultures where one gender (usually male) is preferred over the other, widespread gender selection could easily lead to a dangerous imbalance in the male/female ratio, thereby threatening the future social stability of the culture. In addition, choosing one gender over another is one more form of discrimination that doesn't need to exist in a world where discrimination of many sorts has already caused great havoc.[64] May we never hear parents in our technologically advancing society say to their child, "I knew we should have opted for a girl when we had the chance!" Sadly, such emotional abuse is all too likely to occur in a culture where people increasingly expect technology to fulfill their desires.

21. Should I think differently about genetic intervention to correct medical problems and genetic "design" or "enhancement"?

Many people are comfortable with using genetic intervention to avoid or combat the effects of genetic disease, but far fewer favor the indiscriminate use of genetic technology to make non-medical modifications (commonly referred to as "design" or "enhancement"). An April 2000 Harris Poll indicated that while 71 percent of those surveyed supported the use of genetic intervention to combat genetic defects and 69 percent favored using genetic technology to eliminate physical disabilities, 82 percent were opposed to the prospect of genetically selecting a baby's eye or hair color. The majority of people would, if given the opportunity today, probably choose *not* to design or enhance genetically a particular trait.

This attitude may change, though, as the public's familiarity with genetic technology increases. A major obstacle will then likely be to convince consumers (especially Americans) that their freedom to choose the best advantages for their children should—at times and for the benefit of society—be restricted. In his book *Remaking Eden,* Princeton molecular biologist Lee Silver points out that attempts to limit medical (including genetic) interventions to the treating and curing of disease would likely fail in American society. After all, allowing people to use medical services and technologies (even for non-medical purposes such as facelifts and "tummy tucks") as long as they can afford it is the American way.[65]

While you may disagree with such an unrestricted use of medical technology, Silver quite accurately characterizes the attitude of many and the guiding philosophy of our times. Such a mindset will not permit limitations on genetic design/enhancement unless sound arguments can be presented that the disadvantages outweigh the advantages. Even then, the values of personal choice and commercialism are so prevalent and powerful in free societies that the prevention of genetic design/enhancement technologies may not occur unless people experience their effects to be devastating. Moreover, if parents believe that their children may stand to benefit from genetic enhancement, denying them this opportunity for any reason may become exceedingly difficult.

Arriving at a fair and consistent ethical system for assessing genetic intervention will be more difficult than it seems. It's tempting to suggest that we should simply employ genetic technology for strictly medical reasons and thereby avoid entirely the pitfalls of "designer babies." Such a system would likely fail, however, as the line between these two types of intervention might not always be clear. For example, professionals who work with learning disabled individuals routinely regard learning disabilities as medical problems. To correct these problems will (if all goes well) result in the "enhancement" of traits such as IQ, verbal skill, and memory. If a person's disability is only mild or moderate, the question as to whether a particular genetic intervention is being performed for "medical" or "enhancement" purposes will likely become increasingly complex. As author Robert Wright has pointed out, "At some point you cross the line between handicap and inconvenience, but people will disagree about where."[66]

The biblical mandate for genetic intervention appears to connect much more directly and firmly to healing than to enhancing. John Feinberg suggests that since suffering and death are realities in the world because of sin—and Scripture teaches that we are to fight sin and its consequences—we should use genetic technology as a tool to fight against those realities.[67] But determining which traits and other realities ultimately result from sin and are appropriate to be combated genetically will hardly be simple.

22. Is it okay to design or enhance genetically a person's physical characteristics and/or mental abilities?

Outside of a genetic context, we frequently seek to enhance personal characteristics such as appearance (e.g., we brush our hair) and mental ability (e.g., we pursue education in order to equip ourselves for various tasks and professions). The answer to this question will explore in greater detail whether something intrinsic to genetic design/enhancement itself raises the stakes and should, therefore, give us greater pause.

In 1999, a group of researchers from Princeton, MIT, and Washington University startled the world with an amazing announcement. They claimed that with the simple addition of a single gene, they created mice that were more curious, learned faster, and remembered longer than mice that were not genetically engineered.[68] While Joe Tsien, the group's lead scientist, stated that he had no plans to apply his discovery to humans, the same gene exists in people and probably performs an identical function.

What if we *could* enhance memory in humans by adding a single gene? What could possibly be the downside? Tsien's mice appeared normal in all other respects; however, they may have an increased susceptibility to strokes since the protein made from the added gene also plays a role in whether or not one suffers a stroke. This same protein also plays a role in sensitizing the brain to drugs like cocaine, heroin, and amphetamines and may also act as a trigger for chronic pain. Furthermore, forgetting traumatic events can be tremendously therapeutic. What struggles might an intellectually enhanced individual have if he or she is unable to forget?[69] Since memory is obviously much more complex in humans than it is in mice, studying the effects of increased memory in mice would not provide us with a sufficient answer to this question.

A host of significant problems looms on the horizon should genetic design/enhancement of human beings be broadly permitted. Who, for example, will pay for these interventions? If the consumer alone pays for such enhancements, then only those who are rich will be able to afford them. Such an arrangement would likely result in the creation of two separate classes of people—the *enhanced* and the *non-enhanced*. The movie *Gattaca* offers a disturbing glimpse of such a society and the discrimination that it spawns.

Another significant problem is that choosing genetic design/enhancement therapies for children will often constitute irrevocable consequences that they would not have chosen. These consequences may become particularly problematic when one considers the possibility of multiple effects stemming from the same gene (as arose in Joe Tsien's mice study). For example, what if enhancing verbal skills reduces athletic abilities? Also, could a child sue his or her parents if an enhancement therapy results in a trait (e.g., musical talent) that in the future becomes less desirable than another characteristic (e.g., artistic skill)? Such a circumstance is not at all inconceivable in a world where some individuals are already suing physicians for "wrongful life" (a life whose quality is so poor that it [the life] wrongs the one living it).

Further, enhanced individuals would almost certainly face the pressure of meeting exaggerated expectations. The normally produced child of a genius faces pressure enough, but what if the child of a genius has his or her intellectual gifts genetically boosted? Shouldn't the child's intellect then exceed the parent's? The assumption is yes. Such an expectation is, though, inappropriate and would almost certainly be damaging for the child. Even children in our "pre-enhancement" era often face unfair expectations based on who their parents are, or feel pressure to satisfy the desires of parents whose own dreams were never fulfilled. Such expectations will almost certainly be greater if parents have *intentionally* altered their children's genes to produce desired effects.

Included in the myriad of concerns raised by genetic design/enhancement is the issue of *consent*. While certain types of enhancements may seem appealing, not everyone would wish to be genetically enhanced. Some parents greatly value intelligence and desire that their child attend an Ivy League school. Thus, they might impose their values and standards upon their child as an embryo by genetically enhancing him or her to be extraordinarily intelligent. Such an imposition of desires on a non-consenting individual would open the door to excessive parental control and perhaps even governmental tyranny of the type witnessed in Nazi Germany (see answer to question 23). Even if consent is given, it may not be truly informed since the implications of particular enhancements would likely not be fully realized or understood.

Immense complexities and difficulties, then, surround the issue of genetic design/enhancement. Some interventions could result in harmful consequences, while others may not. Some effects will be immediate, while others may not be fully apparent for generations to come. Since the potential problems and ethical issues are so pervasive, it seems prudent that this type of research in humans should not—at least for now—be done.

23. Could genetic engineering lead to the creation of a "super race"?

This question usually alludes to the Nazi regime of the late 1930s and early 1940s. While the regime's genocide and "race purification" programs were certainly criminal in nature, many people do not realize that they were rooted in the emerging science of the time. Not immune to the lure of these new philosophies, America began to participate in "eugenic" experiments in the late nineteenth and early twentieth centuries. The *eugenics movement* was driven primarily by developments in genetics, which suggested that "human progress" could be advanced through *selective breeding*. In the United States, several state laws were proposed that advocated sterilization of the "unfit" (e.g., the feebleminded, mentally ill, and repeat criminals).[70] In each case, the main proponents for forced sterilization came from established leaders in the scientific, medical, and mental health communities.

The motivation for such policies was to remove harmful and debilitating genes from the population. Between 1907 and 1937, thirty-two states in America passed compulsory sterilization laws, some even citing Nazi Germany's "bold experiment in mass sterilization."[71] The United States Supreme Court voiced its approval in the 1927 *Buck vs. Bell* decision, in which Oliver Wendell Holmes made his infamous statement, "Three generations of imbeciles are enough."[72] Germany passed its own sterilization law in 1933 (sterilizing over forty thousand people by 1937), and over 50 percent of the German medical community eventually joined the Nazi party.[73] We should be sobered today by the fact that the main push for eugenics, forced sterilization, and euthanasia came from the escalating costs of medical care.

The recognition that genes do not wholly shape who we are (see answer to question 4) dealt compulsory eugenics a severe blow. By the early 1950s, the few sterilization laws that still existed were not upheld. Today, many European countries retain strict laws against forced eugenics. But future pressure either to resist or support eugenic ideas will likely not come primarily from governments. Some from within the scientific community are now calling for eugenics of a different kind—a eugenics based on a greater knowledge of genetics and on the demands of a society devoted to personal choice (a value reflected in and encouraged by the U.S. Supreme Court's *Roe vs. Wade* abortion decision).[74]

The precise genetic knowledge and techniques necessary to create a "super race" are, however, currently lacking. We don't yet know the nature and extent of the role that genes play in determining popularly desired traits such as athletic ability, beauty, intelligence, and personality. As our knowledge of human genetics grows, we may learn that most forms of genetic enhancement simply cannot be done. We already know, for example, that very few genes have only a single function—most genes have multiple effects, which are sometimes expressed in different tissues. Thus, a gene engineered to enhance a particular trait may inadvertently malfunction in a different tissue. Moreover, specific traits may be affected by a host of genes. In mice, for example, genes contributing to the texture of hair and skin are known to be located on at least seventy-two chromosome sites. Similarly, skin color in humans is primarily determined by a large group of genes, with additional genes likely to have smaller, less noticeable color effects. Changing just one gene may therefore have a rather negligible effect on a particular trait.[75]

Furthermore, no evidence exists to suggest that enhancing traits such as athletic ability, beauty, intelligence, and personality—should we gain the capacity to do this— would result in an "improved" human being. History has, in fact, demonstrated to the contrary that such heightened abilities and characteristics tend to foster arrogance, narcissism, and pride. Genetically enhanced traits might simply serve to magnify our own sinful weaknesses, an outcome that wouldn't be very "super" at all.

What will happen if the information we learn about various genes *does,* in fact, permit the engineering of desirable traits? Although government coordination of a new forced eugenics is unlikely, an endorsement of eugenics, rooted in our current worldview, has already crept in the door. Molecular biologist Lee Silver believes that the combination of new genetic technologies and the American philosophy of personal freedom and autonomy will, in the next two hundred to three hundred years, spawn two very distinct classes of people: (1) the genetically enhanced, or "GenRich," and (2) the "Naturals," those who have not been genetically enhanced. Since it's unlikely that these two groups would interbreed (why dilute your enhanced genes?), the possibility of establishing separate classes looms large.[76] Silver admits that he's going out on a limb with his prediction, as many scientific and ethical hurdles must be overcome in order for his "class system" to become a reality. Whether or not this undesirable scenario unfolds will depend largely upon the decisions of individual couples. For this reason, Christians and other like-minded people need to join the debate and help ensure that genetic technology is used to liberate and heal, rather than to discriminate against and otherwise harm people.

24. Are we "playing God" by tampering with genes?

The concept of "playing God" means different things to different people.[77] Whenever this phrase is used in casual conversation, we should ask what is meant by it. For some it may actually have nothing to do with God at all, but may simply be used as an expression to convey a sense of wonder and concern regarding the power that human beings wield over nature. While such power should certainly be exercised with caution, stifling scientific research out of fear is never the way to proceed. Doing so is, ironically, as much an improper exercise of authority as is blindly supporting all such research. As Christians, we must argue strongly against unreflective opposition to, as well as unbridled support of, genetic intervention.

For some Christians, however, "playing God" means performing tasks that are reserved for God and God alone. If this is what manipulation of our genetic makeup entails, then concerns about "playing God" are justified. Such a view often incorrectly assumes, however, that because God makes decisions and acts in areas which we have previously not engaged, such realms are therefore forever off limits to human beings. This view reflects an understandable fear of delving into the unknown and of the possible harmful consequences of doing so. People fear what we might learn about ourselves, what new responsibilities genetic knowledge will place on our shoulders, and that such knowledge will be used for harm instead of good. Genetic knowledge in itself is not evil, however, nor are all genetic technologies. People are primarily responsible for the outcomes of genetic intervention (see answer to question 8), and they should seek to further, rather than hinder, God's purposes in this domain.

Because human beings have been made in God's image, we, in a sense, "play God" by imitating Him. Our works of art, construction of buildings, and management of natural parks are just some of the ways we imitate God for the good of His creation. We have, of course, also long engaged in genetic alteration of God's handiwork. Our current varieties of corn, wheat, flowers, cattle, dogs, and horses, for example, bear little resemblance to the original stock in nature. Through selective breeding for our own benefit, we've manipulated these species over the millennia. The question then becomes whether or not seeking to alter life that was created by God, though marred by the Fall, is an appropriate exercise of human stewardship and creativity.

If we are to "play God" properly by imitating Him as

images (reflections) of His goodness, and as responsible stewards over all that He has made, then we need to study His creation to protect and preserve it and apply what we learn with humility and compassion. In addition to recent advances in genetics, scientific breakthroughs in *cloning* (see answers to questions 35–40), *stem cell research* (see answers to questions 32–34), *transgenics* (see answers to questions 28–29), and *xenotransplantation* (see answer to question 30), the nature, power, and specificity of our creative abilities have greatly increased. Such abilities should be used in an ethical manner to improve the lives of human beings struggling in a fallen world.

25. Does the law allow researchers to patent the genes they identify and engineer?

Before 1980, there was no confusion on this matter in the United States. The U.S. Patent and Trademark Office (PTO) refused to grant *patents* for living organisms because they were "products of nature."[78] This changed in 1980, however, when the U.S. Supreme Court allowed a patent to be issued to General Electric (GE) for a bacterial strain that the company had engineered to digest petroleum from oil spills.[79] The court ruled that since scientists had manipulated the organisms, they were no longer "products of nature." Then, in 1988, the PTO granted a patent for genetically engineered oysters by arguing that they, too, were not natural products.[80] In that same year, Harvard University was granted a patent for a genetically altered mouse used in cancer research. Six years later, Harvard again patented a mouse (also for use in cancer research) that had been genetically engineered to have an enlarged prostate gland.

The PTO has traditionally stated that human beings cannot be patented, maintaining that such an act would convey ownership of a human being and thus be contrary to the Thirteenth Amendment. The same does not hold true, though, for human body parts. In 1984, a patent was granted to the University of California for cells cultured from a diseased human spleen. A challenge to this patent came from the individual whose spleen was the source of the cells, but his case was rejected on the grounds that he did not own the cells.[81] Even the prohibition against patenting whole human beings was called into question when an animal cloning patent that is apparently applicable to human cloning—as well as to the *products* of human cloning—was granted by the PTO on April 3, 2001.[82]

Thousands of plant, animal, and human genes have already been patented, with many more patents still pending. Scientists are increasingly endeavoring to protect their research investments by applying for patents on identified gene sequences, sometimes even before their medical utility is known. In response, the PTO has proposed a clarification of its regulations concerning gene patents: a patent will be awarded only when the function of the protein coded for by the gene is known.[83] But thousands of gene patents have been applied for even though this criterion is not met, and the PTO regulations don't go far enough in that empirical demonstration of the protein's function is not required.

Can scientists legally patent the genes they identify and create? The simple answer to this question is "yes."

26. Should researchers be allowed to patent the genes they identify and engineer?

The purpose for obtaining a *patent* is to give the applicant sole possession of all rights and privileges that accompany the use of a newly discovered or engineered product. Pharmaceutical firms patent genes to cover the cost of the extensive research required to produce safe, effective gene therapies. Without the patents, drug companies could not prevent others from unjustly cashing in on medical advances. Thus, they could not attract investors, who supply the funds necessary to support both the research and the long-term process of gaining federal approval for new genetically derived therapies.

Many people in the scientific community are leery, however, about the government issuing patents for actual gene sequences. These researchers fear that the license fees they must pay to the patent owner(s) for use of a particular sequence will drive up the costs of their own experimental work. Overlapping and competing claims may stifle innovative research, ultimately leading to greater health care costs for the consumer.

In addition to the above implications of gene patenting, C. Ben Mitchell, Ph.D., Senior Fellow of The Center for Bioethics and Human Dignity, has noted that gene patents have raised six social justice concerns that require serious consideration. First, such patents make important medical products more expensive and less accessible to patients. Second, gene patents promote secrecy among scientists and therefore hinder the exchange of information. Third, these patents exploit taxpayer-supported research by allowing government-funded university laboratories to profit from their patented research discoveries. Fourth, gene patents promote unsustainable and inequitable agricultural policies in that patenting may lead to the proliferation of genetically improved plants and animals—which may in turn lead farmers to overly depend upon them, thereby reducing necessary genetic diversity. Fifth, First World patenting of Third World genetic resources constitutes theft of community resources. Sixth, patents on living organisms are morally objectionable to many.[84]

Furthermore, theological concerns exist that should, when combined with the above objections, motivate a call for at least a moratorium on patenting human and animal life and the components thereof. Even though scientists may genetically alter an organism, body part, cell line, or gene, this does not change the simple truth that God, not human beings, is the Creator of life. While it is appropriate, then, to patent the processes utilized in biological discovery, it would seem that the biological entities themselves should remain free of restrictions.

27. Since there are so many unknown risks, should we oppose genetic engineering in general?

The risks posed by gene therapy (see answers to questions 16 and 17), the creation of transgenic animals (see answers to questions 28 and 29), xenotransplantation (see answer to question 30), and other forms of genetic intervention are real and should not be taken lightly. Genetic engineering can, however, offer great benefits that should be pursued within wise and ethical guidelines. The current state of genetic research, coupled with substantial public support and government and private funding, clearly show that we have already taken significant steps down the path of genetic intervention. Prohibiting all forms of genetic engineering is neither desirable nor possible, therefore we should commit to positively influencing how our increasing knowledge of the human genome is applied.

Genetic engineering should proceed at a cautious and prudent pace, allowing for public comment and direction. Such input will, hopefully, increase the likelihood of achieving the many benefits of genetic technology without desecrating human life and dignity in the process. If, as Christians, we take seriously our role as stewards of God's creation, we should explore opportunities to counter the negative effects of sin on that which He has created. We should also acknowledge, however, that the worthy and honorable quest to better people's lives must be pursued with and tempered by a keen awareness of humanity's propensity to serve self at any cost.

Christians have an opportunity to help foster an appropriate and compassionate use of genetic technology and to make the wisdom of the gospel attractive and relevant to a lost and hurting generation. Such cannot happen, however, if we remain uninformed about genetic issues or withdraw from discussions in the public arena. Churches and Christian institutions of higher learning must seek to equip people to stay abreast of the developments in genetic research and to communicate to a pluralistic society the Christian worldview and its corresponding concerns. This task will not be an easy one, but it is absolutely necessary.

28. Is transgenics—combining genes from different species—okay?

In the past, human beings have created many "new" creatures. Mules, dogs, cattle, chickens, wheat, and corn would not exist in their present forms and varieties apart from human intervention. Such intervention was often carried out in a very crude way by breeding members of different species with the hope that their offspring would exhibit desired traits. This method promised little to no certainty of success, as most of these experiments failed to achieve the intended outcome. *Transgenics* is a more targeted method of creating "new" organisms by adding genetic material from one species to the genome of another species.

If the mere idea of adding something foreign to an organism is troublesome, consider that we do this whenever we take, for example, medications. Our bodies would never have access to most of these substances in nature. Still, the prospect of transferring a gene or genes from one species into another species certainly raises new issues that demand our consideration. It is reasonable to ask if we have the wisdom even to attempt such gene transfers, in part because their impact is potentially so much more enduring and difficult to undo than that of traditional medicines and vitamins.

To answer this question, we should first identify the nature and purpose of gene transfers. We need to ask whether transgenic alteration may inappropriately cross species barriers that God instituted at creation.[85] In making this determination, it's important to recognize that a single gene does not define a species. Bacteria are composed of thousands of genes, and it's been estimated that each human being possesses as many as thirty thousand to forty thousand genes. The collective genome of an organism certainly does define its biological identity to a significant degree, but an individual gene does not carry that level of power. Therefore, transferring just a single gene from one organism to another does not create a *hybrid* (genetic combination of two species) in the traditional sense. It is therefore an overstatement to say that by transferring a gene from one species to another species, we've created a new organism and have certainly gone beyond the boundaries of imitating God's creative activity. To the contrary, such limited gene transfer would, generally speaking, seem to be acceptable if carried out for good purposes.

On the other hand, Scripture tells us that God designed plants, animals, and humans in a way that would allow them to reproduce after their own kind or seed (Gen. 1:11–12). Should Christians then object to the creation of organisms with genes from more than one species? A careful

71

reading of Leviticus suggests that God is concerned not just with the physical defilement resulting from a human being mating with an animal, but also with the nature of the offspring that might result. It would thus seem that we should unquestionably reject the complete fusion of human and animal genomes, e.g., combining a human egg with a horse's sperm (though many such fusions would likely not be possible).

We do not know at what point transferring genetic material from a member of one species (e.g., a human being) to another (e.g., a rabbit) would result in an organism with features of both species (e.g., a "human-like rabbit"), so it would seem wisest to refrain from such an endeavor (see answer to question 29).

Transgenic technology provides a power never before possessed by human beings. We now have the ability to design or create a new variety of organism by intentionally altering its genetic structure for a very specific and often very limited purpose. While such purposes can be good, they may also be evil. Scientists might, for example, transgenically modify the genome of a known pathogenic, or even deadly, bacterium with the intent of creating a biological weapon designed to unleash non-treatable human illnesses. Such a capability is sobering and should cause us to pause and reflect seriously on the potential negative implications of this technology. It's also possible for transgenic organisms created for agricultural and medical purposes to develop in ways that were not intended or foreseen.

Given such scenarios, it is imperative that proper and extensive tests be performed and sufficient knowledge be gained to ensure (to the greatest extent possible) that no unnecessary harm will come to human beings, animals, or the environment as a result of transgenic modification. If we are to employ this technology in ethical and responsible ways, care must also be taken throughout the entire research process to ensure that the dignity of human beings is not violated and that the integrity of other species is not grossly undermined.

29. Is it wrong to place human genes in an animal?

It is important to remember that a gene is simply a string of DNA, which is a molecule (see answer to question 1). Transferring a gene from one organism to another does not, therefore, necessarily create an ethical dilemma. However, a new dimension is added to this question when the genes involved in the transfer are human. Human beings are unique in God's creation, as we alone are made in His image and likeness. It is important to recognize, however, that the image of God is not wholly contained in our DNA, as DNA is just a molecule. Certainly, the human physical body is designed to join with the soul and spirit to constitute a unified person made in God's image, but the physical body alone is not the bearer of the image itself. Otherwise, corpses would need to receive fundamentally the same treatment and protections as living human beings.

Already, human proteins necessary for treating symptoms in patients suffering from hemophilia, cystic fibrosis, cancer, diabetes, and many other diseases are being produced from transgenic animals that have had human DNA inserted into their genomes (see answer to question 9).[86] Transgenic pigs, for example, are being engineered to produce Factor VIII, a blood-clotting agent needed by certain hemophiliacs (who can easily bleed to death if injured because their blood does not clot as it should). The necessary protein is produced in the pigs as part of their milk, from which it can be extracted and sold as a much-needed drug that is less expensive than and superior in quality to those produced by other methods.[87] Mice, rats, pigs, and sheep are also being engineered with specific human genes to serve as models in testing new therapies for human diseases.

In April 2000, BioTransplant, Inc., was issued a patent for their technique of engineering pigs that carried human HLA genes. The technique was developed for the purpose of reducing the risk that transplant patients would reject porcine-derived cells, tissues, and organs. These transgenic pigs were not changed in any substantive way—they're still pigs—and will likely live normal, healthy lives. Although additional advances must be made before pig organs can be transplanted into people, achievements like these are believed by many to be a major step toward solving the current shortage of transplantable human organs. As such, this type of transgenic research is a worthy and permissible endeavor that should be encouraged, albeit with the necessary safeguards.[88]

On the other hand, some scientists conducting research in human/animal transgenics may have goals that are not so praiseworthy. Some scientists, for example, have suggested that we begin transferring human genes into gorillas and chimpanzees for the purpose of deciphering which genes are responsible for the distinction between human beings and the great apes. If the agenda behind such research is to blur the line between these species, then it should not be encouraged. Some animal rights activists have already requested that some "human" rights be bestowed upon chimps and gorillas.[89] After all, they say, humans and apes have as much as 99 percent of their genetic material in common.[90] As mentioned previously, however, human beings are distinct by virtue of their creation in the image of God—an image that involves far more than genetic makeup. If such research proceeds, it will be interesting to see—should it be discovered that relatively few genetic differences exist between these species—whether people will acknowledge that the outwardly observable variations must be due to non-genetic factors, or if people will instead claim that there are few significant distinctions after all.

Regardless of the stated goal of transgenics, it would seem wise to establish limitations on the amount of human genetic material that can be transferred to another species. While it's true that our DNA does not wholly contain the image of God, we should nevertheless be concerned about the possibility of non-human organisms exhibiting human-like traits and/or behavior. Although early transgenic organisms generally contained no more than two genes from another species, scientists have begun to insert increasingly larger amounts of human DNA (even fractions of entire chromosomes) into animals of various species. Such laboratory practice raises concerns that a transgenic animal might exhibit uniquely human characteristics—a prospect that should be avoided.

30. Is xenotransplantation—transplanting animal organs into humans—okay?

In Genesis 1:21, God Himself made clothes from animal skins for Adam and Eve. Later, He gave specific permission to use animals for food after He went to great lengths to preserve them from the Flood (Gen. 9:3). Human beings have used domesticated animals for their own purposes for millennia. For example, before genetic engineering allowed human insulin to be produced from bacteria, the spleens of slaughterhouse pigs were routinely harvested to extract insulin for human diabetics (see answer to question 7). This does not mean, however, that human beings are to have no regard whatsoever for animal welfare. To the contrary, research on animals must be justifiable and must occur in accordance with existing standards for experimentation on non-human species.

As mentioned in the answer to question 29, scientists are currently working on transferring human genes into pigs so that the pig organs will not be rejected when transplanted into human patients. Though this would eliminate the cross-species rejection problem, the problem of individual rejection (much the same as that which sometimes typifies transplants between humans) would remain. Another strategy for avoiding rejection is to inactivate, or "knock out," the genes in animals that are responsible for triggering rejection in other species. In March 2000, PPL Therapeutics, the same company that cloned Dolly the sheep, moved closer to the goal of using pig organs for human transplantation by producing the world's first cloned pigs.[91] They did so as a first step in their endeavor to produce mass numbers of pigs genetically engineered to lack the genes that trigger rejection in humans. In January 2002, PPL scientists, in addition to researchers at the University of Missouri and Immerge BioTherapeutics, announced that they had succeeded in cloning pigs that they had genetically engineered to lack one copy of the gene responsible for the production of a sugar molecule that plays a key role in triggering rejection.[92] The next goal was to produce cloned pigs lacking *both* copies of the gene— a feat that was accomplished by PPL in August of 2002. In January 2003, the University of Missouri and Immerge BioTherapeutics followed suit, announcing that they too had produced "double knock-out" cloned pigs.[93] In February 2003, the Wisconsin-based company Infigen, Inc., also reported having achieved the same success in collaboration with Immerge BioTherapeutics.[94]

Building on the transgenic approach, researchers at the University of Milan announced in October 2002 that they had mixed pig sperm with a human gene and then used the altered sperm to fertilize pig eggs in hopes of one day producing pigs whose organs would pose no threat of rejection to humans. The procedure is much simpler and more efficient than standard transgenic techniques, although it is predicted that five to seven other pig genes will need to be rendered inactive or replaced by human genes before successful transplantation can occur.[95]

Although the prospect of saving human lives via the transplantation of animal organs is certainly exciting and the above successes are promising, a myriad of concerns remain. A major fear regarding xenotransplantation is the possibility of transmission of disease from animals to humans. Some people worry, for example, that transplanting a pig organ into a human patient might enable viruses such as porcine endogenous retrovirus (PERV) to mutate and become capable of causing disease in humans.[96] In response to this concern, the American Medical Association (AMA) has proposed ethical guidelines to govern xenotransplantation. Among these is the requirement that patients receiving animal organs first agree to undergo regular medical evaluations so that the risk of a public health threat can be reduced.

Other concerns include the psychological reaction (or "yuk factor" response) of many to the prospect of placing animal organs into humans. How might recipients of such organs respond and be responded to by others? A further issue raised by this technology is the question of whether animals have rights or interests that should prohibit their sacrifice as a means of improving or saving human life. Finally, some may object that xenotransplantation violates fundamental species lines established by God.[97]

While we believe that the above concerns merit serious consideration, many critically ill patients would stand to benefit if animal organs prove to be a solution to the current shortage of transplantable human organs. In addition to offering a great number of people a chance at life, using animal organs for transplantation would eliminate the unpleasant current reality that another person must die if prospective organ recipients are to live. It would thus seem that research directed toward the use of animal organs for transplant should be pursued, although extreme care should be taken to ensure that the pursuit is carried out in an ethical and safe manner.

<hr>

STEM CELL RESEARCH AND CLONING

31. What is an embryo?

A proper regard for the *human embryo* is essential in grappling with matters such as stem cell research, cloning, and certain forms of genetic intervention. However, confusion over the precise meaning of the term "human embryo" can lead to disagreements on these issues. It is therefore important to have as clear an understanding as possible regarding the nature of the human embryo before determinations of how embryos should be treated are made.

In the mid-1800s, the German anatomist Ernst Haeckel characterized the human embryo as a mere cell consisting of "homogeneous globules of plasm."[98] In this period of time, before the advent of molecular biology, scientists were not privy to the magnificent complexity of embryonic human life. Those who wished to value human embryos often had to appeal to religion or philosophy, rather than scientific fact, in part because the intricate genetic makeup of the human species—already present in these very young members of the human family—was not yet known to researchers.[99] This began to change, however, as insights into the mechanism of conception and the discovery of the nature, structure, and role of DNA (as well as a myriad of technological advances) allowed researchers to peer ever more closely into the human embryo.

As a result of this increasing scientific sophistication, both secular and religious institutions alike now acknowledge that the human embryo exists at an individual's very early stages of development. Since 1996, the U.S. Congress has defined the extracorporeal human embryo as an "organism . . . that is derived by fertilization, parthenogenesis,[100] cloning, or any other means from one or more human gametes [eggs and sperm] or human diploid cells [cells having forty-six chromosomes, the normal number in human beings]."[101] In referring to the *extracorporeal* human embryo (an embryo existing outside of a woman's body due to having been created through in vitro fertilization, etc.), Congress recognizes that a human embryo is *what* it is no matter *where* it is.

An existing Congressional ban forbids federal funding for any research in which human embryos are destroyed or subjected to risk of harm.[102] Yet even governmental bodies that support the destruction of embryos for research purposes (such as the National Bioethics Advisory Commission [NBAC] and the National Institutes of Health [NIH]) agree that a human life is present at the time of fertilization.[103] Finally, the National Academy of Sciences (NAS) defines the human embryo as a "developing human from fertilization . . . until the end of the eighth week of gestation."[104] Thus, labels that refer to the embryo as a "pre-embryo" or "potential human being" are scientifically invalid; this early life is presently (not merely soon-to-be) an embryo and is actually (not merely potentially) human. Consequently, we should reject terms that seek to detract from the moral status and dignity of early human life and/ or to desensitize us to the destruction of that life.[105]

Scientific and scriptural evidence together—both genetics and Genesis—point to divine involvement in the creation of human beings even at the embryonic stage. And because human embryos are created in God's image, various passages of Scripture affirm and reflect on the preciousness of embryonic human life in the womb.

Psalm 139, for instance, opens with six verses on the vastness of God: God knows everything and is everywhere. The next six verses describe the natural human bent to try to escape from God—to find some context in which we can live our lives however we desire. The psalmist then brings home the fact that there is no such setting—even in the womb, which early on houses an "unformed body" (some translations simply translate this expression as "embryo"). *Nothing* is hidden from God. When we do something "in the womb" (i.e., to the embryo), we are not in a realm of "private choice" or science in which we are welcome to do with an unborn human being whatever we think will benefit us. The embryo is created in the very image of God, and God is watching in the womb (not to mention in the more visible lab) to see how we are treating that image bearer.

God also has another major connection with the embryo. It is through the Incarnation, when God became a human being in the person of Jesus Christ. If only adults are human beings, or one must be born (or at least a well-developed fetus) to be a human being, then we might expect to see God becoming only such. It is hard to imagine Him taking on some form that could have been ethically discarded as a mere blob of material.[106]

32. What is stem cell research?

To answer this question, we must first define the term *stem cell*. The name for this type of cell may be traced to the stem of a plant. The different branches, leaves, and flowers on a plant all originate from the stem. Therefore, a "stem cell" is a type of cell from which other cells have their origin. In a human embryo, stem cells play an important role in the initial formation of many different kinds of tissues. In more mature human beings, stem cells give rise to new cells in order to replace old cells that have died or that no longer function properly.

Some stem cells (those found in early embryos) are naturally capable of generating the more than two hundred types of tissue in the human body. Other stem cells (such as "mesenchymal" stem cells) have long been known for their ability to produce several kinds of tissue (such as bone, cartilage, or fat). Still other types of stem cells have been thought to have the capacity to give rise to cells of only a single kind (e.g., a liver cell that can produce only more liver cells and not cells with other functions).

Given the startling advances in animal cloning (see answer to question 36), it's tempting to think that just about any cell can be stimulated to give rise to all human tissue types. To the extent that human cloning can be carried out successfully (see answer to question 37), such thinking is justified. Cloning involves placing the nuclear genetic material of an adult cell into an egg cell that has been stripped of its nucleus. The egg is then "tricked" into acting as if it had been fertilized, with the result that a new embryo—containing stem cells capable of generating all types of human tissue during its development—is created.

The current debate over stem cell research has been raging since two teams of U.S. scientists announced in November of 1998 that they had successfully isolated and cultured human embryonic stem cells in the laboratory.[107] This feat was a scientific first and set off high hopes in patients and researchers alike that these stem cells held the key to treating—and even curing—many of the most devastating human afflictions.[108] The basis of this belief rests with the nature of an embryo's stem cells: since they are capable of generating all types of human tissue, having access to them in the lab would potentially provide scientists with the means to produce new heart tissue, liver tissue, brain tissue, lung tissue, or any other type of tissue that a patient might need.

The stem cell batches that sparked the ongoing ethical debate did so because they were derived from "surplus" human embryos (obtained from fertility clinics) who were destroyed in the process, as well as from fetal tissue obtained from electively aborted babies. Availability of these types of stem cells (referred to as *embryonic* or *fetal* stem cells) is primarily dependent upon decisions to intentionally end the life of very young human beings. Proponents of research on such cells say that the research should proceed because of its allegedly great promise for medical benefit, while opponents deem it unethical—regardless of what benefits might emerge.

A second type of stem cell is that which is not derived from a human embryo or fetus. This type of stem cell is commonly referred to as an *adult* stem cell—even though such cells are also found in children and teenagers and may even be collected from the umbilical cord of a newborn infant. These cells do not carry the same ethical concerns as do embryonic stem cells, since the individuals from whom the cells are obtained are not harmed in any way. Of great interest to scientists, patients, and much of the general public is the emergence of unexpected evidence that adult stem cells also have a great capacity to develop into many (and perhaps all) types of human tissue. Such cells may prove equivalent—or even superior—to embryonic stem cells in terms of their therapeutic value (see answer to question 33). This is very good news, since the destruction of human embryos required to obtain embryonic stem cells is unethical.

33. Is "adult" stem cell research as promising as embryonic stem cell research?

Embryonic stem cell research has received the majority of media hype and public attention. The obstacles encountered in this research, however—combined with the unprecedented advances reported by scientists working on "adult" stem cells—may actually serve to underscore the therapeutic potential of adult stem cell research as being superior to embryonic.

When embryonic stem cells were first isolated and cultured in the laboratory in November 1998, scientists heralded the cells' ability to form all types of human tissue as holding the key to the development of revolutionary medical therapies for a whole host of human illnesses.[109] This research has been fraught, however, with serious challenges that continue to plague scientists. For example, in order for a single stem cell to develop into a desired tissue type, exposure to specific cellular cues may be required. Stem cell researchers have thus far been unable to decipher these cues. Another difficulty encountered by scientists is the threat of rejection posed by embryonic stem cells. If an embryo's stem cells are not immunologically compatible with the patient into whom they are injected, rejection might occur as would be seen if the tissues of an organ donor and recipient were incompatible. (This has caused many proponents of embryonic stem cell research to speak out in favor of producing cloned embryos for the purpose of obtaining their stem cells—see answers to questions 38 and 39). Furthermore, embryonic stem cells have been associated with the undesirable consequence of tumor formation.[110] Unless scientists learn how to surmount these significant obstacles, the alleged promise surrounding embryonic stem cell research may never be realized.

The scientific developments encountered in *adult stem cell research,* on the other hand, have been much more favorable. While it was once believed that only embryonic stem cells had the capacity to develop into all types of tissue in the human body, adult stem cells have been demonstrated to be far more pliable than initially believed. Rather than an adult neural stem cell being limited to making only more neural cells, for example, such cells have reportedly been demonstrated in mice to produce bone marrow cells that give rise to blood.[111] Among the many other exciting advances in adult stem cell research is the discovery of adult bone marrow stem cells (known as "multipotent adult progenitor cells," or "MAPCs") that apparently can develop into virtually any human tissue type and reproduce extensively in culture without losing their broad developmental capacity. Such cells were not shown to form tumors and may be transplanted into a patient from whom they were obtained, thereby avoiding the risk of rejection.[112] Adult stem cells may also in many cases carry the added advantage of requiring less intricate manipulation to induce them to produce the desired tissue.

Stem cells derived from adults have, in fact, already been used to benefit human patients suffering from a number of afflictions—including autoimmune diseases (multiple sclerosis, lupus, juvenile and rheumatoid arthritis), stroke, immunodeficiencies, anemia, Epstein-Barr virus infection, corneal damage (with full vision restored in most patients), blood and liver diseases, osteogenesis imperfecta, various cancers (in conjunction with chemotherapy and/or radiation), heart attack, and cartilage and bone damage. Although much of the public believes that embryonic stem cell research holds more medical promise than does adult stem cell research, no comparative list of treatments exists for the former.[113]

Another exciting potential alternative to embryonic stem cell research is the possibility of inducing adult non-stem cells to convert directly into cells of many different types (e.g., immune system cells, nerve cells, insulin-secreting pancreatic cells) without first having to pass through the embryonic stage (as is the case with cloning). A group of Norwegian scientists have reportedly already succeeded in coaxing an adult skin cell to behave as immune system cells and nerve cells in the laboratory.[114] If such research proves to be clinically applicable, tissue transplants could conceivably be performed as a result of manipulating cells taken directly from a patient—thereby avoiding the many difficulties of immunological rejection, as well as the unethical destruction of embryos.

Adult stem cell research and other alternatives to embryonic stem cell research do, indeed, appear to be incredibly promising and may render the destruction of human embryos wholly unnecessary. Yet certain challenges and obstacles unique to these alternatives may arise. In the event that non-embryonic research fails to prove as clinically useful as current experiments suggest, destroying embryos is, nevertheless, unethical and should not be permitted—regardless of whether other means for improving human health exist.

34. Should I be in favor of stem cell research?

Many scientists and patients believe that stem cell research holds the key to developing treatments or cures for

Alzheimer's, Parkinson's, brain and spinal cord injuries, heart disease, severe burns, and a host of other human diseases and injuries. We applaud the desire to heal people, as long as the means for doing so are ethical. Because *embryonic stem cell research* is (at least currently) dependent upon the destruction of human embryos, such research is unethical—regardless of its potential for improving human health. By contrast, *adult stem cell research*—which does not require the destruction of embryos—is an ethical means of alleviating suffering and should be enthusiastically encouraged.

The human embryo is a member of the human species and is, therefore, a human being (see answer to question 31). Each of us was once an embryo—full of unrealized potential, but no less a human being.[115] Painful lessons from U.S., as well as world, history illustrate the horrors that result when human beings are used for medical experimentation without their consent. While we share the desire to relieve suffering and restore health, the intentional destruction of some human beings for the alleged good of other human beings is always wrong.[116]

Those who support embryonic stem cell research often do so on the basis that the embryos involved "are going to die anyway." Persons in favor of this research therefore maintain that embryonic sacrifice should be carried out in a manner that will benefit other human beings. The status of such embryos should, however, be regarded as analogous to that of prisoners on death row. Since we do not permit such persons to be killed in order to obtain their organs for life-saving transplants, why should we allow "surplus" human embryos to be destroyed in research on the grounds that they "are going to die anyway"?

Actually, though, "surplus" embryos are *not* necessarily destined to die, as they may be "adopted" and implanted into women who wish to raise them as their own children. Although the embryos would not be genetically related to either the women or their husbands, such adoption would circumvent the high costs of assisted reproduction that are often assumed by couples struggling with infertility. Furthermore, adopting women would be allowed to experience the joys of pregnancy and childbirth. Beyond these benefits, embryonic human beings would be rescued from destruction. Several babies have already been born as a result of embryo adoption, including Mark and Luke Borden whose parents, John and Lucinda, testified against embryonic stem cell research at the July 17, 2001, U.S. House of Representatives hearing on this issue.[117] (For more information on a legally established embryo adoption program, access www.snowflakes.org.)

Most Christians and others who uphold the dignity of embryonic human life agree that the destruction of human embryos for the sake of scientific or medical gain is unethical. Such persons may hold different positions, however, regarding whether it is acceptable to conduct research on stem cells obtained from embryos who have already been destroyed. On August 9, 2001, President Bush announced his decision that federal funds could be used to support only embryonic stem cell research involving embryos who had already been destroyed. Christian and pro-life groups were divided in their reactions. The President believed that his decision was consistent with his earlier campaign promise not to allow federal funding for stem cell research in which embryos are destroyed. By funding research only on cells derived from embryos for whom "the life-and-death decision had already been made," Bush contended that he remained faithful in continuing to oppose the destruction of human embryos. Indeed, the President refused to allot funding for research requiring future embryonic destruction. Some, however, have questioned whether the government can fund *any* embryonic stem cell research without crossing an ethical line. They ask whether the government would be complicit in the evil of destroying human life if they support research that is dependent upon such destruction.[118]

In addressing this dilemma, an important question must be considered: Is it the *destruction of human embryos itself* that should be opposed, or is it *research dependent upon such destruction* that should be condemned? If an embryo has already been destroyed, can a person who objects to such an act support research on the embryo's stem cells and still be regarded as morally upright? Or is supporting the experimentation equivalent to supporting the necessary destruction of embryonic life?[119] The U.S. government is indeed (in some cases) rewarding researchers for destroying human embryos by supplying them with federal funds to experiment upon the embryos' stem cells.[120] As stated above, questions such as these have prompted various responses from the Christian and pro-life communities and clearly demand serious further consideration.

Research on non-embryonic stem cells (so-called "adult" stem cells) appears to be a surprisingly promising and ethically acceptable alternative to embryonic stem cell research (see answer to question 33). Such research is carried out on cells derived not only from adults, but from children, umbilical cords, and cadavers. Adult stem cell research does not require the death or destruction of human beings and should, therefore, be enthusiastically supported.

For more information about the ethical, legal, and scientific aspects of stem cell research, see The Center for Bioethics and Human Dignity's statement[121] and web site (www.cbhd.org) and the Do No Harm web site (www.stemcellresearch.org).

35. What is cloning?

Cloning refers to the making of an identical (or virtually identical) copy of something else. In biotechnology, the earliest clones were of stretches of DNA, and the field of DNA cloning seemed to spring up overnight. Numerous procedures were perfected that allowed researchers to extract a section of DNA from an organism and manipulate it so that literally billions of copies could be manufactured for use in laboratory experiments. Cloning of cells, genes, and other biological material that does not require the destruction of human life may constitute an ethical means of pursuing scientific and medical research. Most of the focus on cloning today, however, pertains to the creation of an organism (typically an animal or human being) with nearly the exact genetic makeup of another organism.

To clone in this sense is to produce a nearly exact genetic copy of a previously existing individual. This reproduction is different than in the case of identical twins. Identical twins are produced when a zygote (single-celled human being originating from the union of a man's sperm and a woman's egg) or an embryo at the four- to eight-cell stage divides to form two distinct individuals. Since both individuals ultimately arise from the same single cell, they share the same genetic material. This accounts for their typically similar appearance and behavior (although differences in both testify that people are much more than their genetics).

Although clones also share virtually all of the same genetic material, the cloning process is wholly different from twinning. If you decided to hire someone to clone yourself, you would be asked to supply some of your cells (e.g., skin cells) to begin the procedure. These cells would be grown in culture for awhile, and then several would be selected and the nucleus of each removed. (The nucleus contains the chromosomes and, therefore, nearly all of your genetic material. Because of the small amount of extra-nuclear mitochondrial DNA, your clone would not be an *exact* genetic replica of yourself.) The nucleus from one of these cells would then be placed in an enucleated egg cell (an egg cell that has had its nucleus removed). Depending upon the procedure being used, the egg cell would be treated (probably electrically stimulated) to cause division as in a normal embryo. The cloned embryo would then either be implanted into a woman—with the hope that the embryo would be brought to term nine months later—or used for research purposes (see answers to questions 38 and 39).

36. Why are plants and animals cloned?

Plants have probably been cloned for millennia. Gardeners seek to create a clone every time they take a cutting from a flower, vegetable, shrub, or tree and place it in water where it sprouts roots, thereby enabling it to grow on its own. The resultant plant is indeed a clone, an exact genetic duplicate of the original plant from which the cutting was taken. Gardeners go through this procedure in order to make more copies of a particular variety of plant that they desire, rather than leave to chance the combination of pollen and egg.

Cloning of animals is quite another matter. Obviously you can't take a simple "cutting" from a pig or sheep and grow another copy! Attempts to clone frogs in the 1950s and 1960s were only partly successful. Frogs cloned from tadpole cells developed into adults,[122] but frogs cloned from the cells of adult frogs grew only to the tadpole stage and then died.[123] Cloning attempts in the 1970s and 1980s seemed to demonstrate to some researchers that mammalian cloning from adult cells was impossible, at least in the near future. The major obstacle to achieving such a feat was believed to be that only certain genes in an adult cell remain functional—e.g., only the genes necessary for lung function are active in a lung cell. Because embryonic or fetal cells have not yet been fully differentiated, the activity of their genes is not similarly limited—thereby accounting for the success in cloning from immature tadpole cells. There seemed to be no way to coerce into becoming active *all* of the genes in an *adult* cell nucleus that were necessary for the successful development of an embryo. This apparent obstacle was seemingly removed when the birth of Dolly the sheep—who was cloned from the cell of an *adult* sheep—was announced in February of 1997.[124]

The original purpose of the sheep cloning experiments was to find a more effective way to reproduce sheep that had been genetically engineered to produce drugs for humans.[125] Sheep and other animals can be genetically engineered to produce a certain human protein or hormone in their milk (see answers to questions 9 and 29). This may be accomplished by taking the human gene for production of this protein or hormone and inserting it into a sheep embryo. It is hoped that the embryo will grow into a sheep with the inserted human gene incorporated into all its cells so that the desired protein is produced. The human protein may then be harvested from the milk and marketed. Such a result is not a certainty, and while the procedure may improve, it likely will never be perfect. Mating a successfully engineered sheep in order to obtain offspring that have the inserted human gene is also not foolproof because the inserted gene is frequently lost. Even mating such a sheep with another genetically engineered sheep containing the same gene may result in lambs that have lost the inserted human gene and are therefore unable to produce the desired protein. Cloning more directly assures that the engineered gene product will not be lost.

On July 24, 1997, the Roslin Institute and PPL Therapeutics (the two institutions that cloned Dolly) announced the birth of five more lambs, this time cloned from fetal cells that carried (and had been proven to express) a human gene for the production of medically important proteins.[126] The lambs' birth marked a significant scientific advance, as they were the first *transgenic* sheep to have originated through cloning (see answers to questions 29 and 30).

In August of 2002, scientists in the U.S. and Japan announced that they had cloned cows that produce human antibodies.[127] Such antibodies—needed to treat patients who have difficulty fighting infections—have traditionally been derived from donated human blood and are therefore in short supply. The researchers hope to increase this supply via cloning cows that produce the antibodies and then infecting the cows with agents known to cause some of the most dreaded human diseases (e.g., AIDS) so that the antibodies produced will be particularly suited to fight these illnesses. Although the cloning achievement appears promising, many obstacles must still be overcome before the antibodies may be used to treat human disease.

Research projects designed to make cloning the family pet a widely accessible option are also underway. The U.S. company Genetic Savings and Clone is currently engaged in a multimillion dollar effort (dubbed the "Missyplicity Project") to clone a beloved mixed-breed pet dog named Missy. While several cloned embryos have been implanted into surrogate dogs, all of them have miscarried for undetermined reasons. Nevertheless, researchers collaborating on the project remain hopeful that they will succeed in cloning Missy. While dog lovers may have to wait a little longer to clone their beloved pets, cat lovers may be able to have their favorite felines cloned much sooner. The world's first cloned cat, "cc" (for "Copycat" or "carbon copy"), was born in December 2001—the creation of scientists at Texas A&M University. The project was underwritten by Genetic Savings and Clone.[128]

Pet cloning is a case of cloning simply for emotional and sentimental reasons. Genetic Savings and Clone receives several inquiries each month from worried pet owners who want to know how much it will cost to have their aging or dying pet cloned. The company responds to such questions and stores frozen tissue samples for those pet owners who wish to proceed—with the hope that clones will eventually be produced. The initial clones are predicted to cost "in the low five figures," with the price becoming lower as efficiency increases.[129]

Regardless of the great love that people have for their pets, the cloning of pets is ethically questionable, as it constitutes a poor use of funds and carries with it false expectations. Furthermore, although a cloned pet would be nearly genetically identical to the animal from which it was cloned, its behaviors and temperament may be quite different from the beloved original.

37. Can human beings be cloned?

When the cloning of Dolly the sheep was announced in 1997 (see answer to question 36), speculation about the possibility of human cloning soon grew. Since humans and sheep are both mammals, many people believed that it was reasonable to assume that the same technique would work with people. Many scientists suggested, however, that the procedure used to create Dolly might not work with humans because of critical differences in early embryological development. When mice were successfully cloned in 1998,[130] though, some scientists asserted that concerns about possible technological barriers to human cloning had suddenly been removed. This opinion arose because mice and humans share many important similarities during the early stages of embryonic cell division; therefore, a technique successful in mice would probably work in human beings. Others, however, still contended that the attempt to clone humans would be met with its own unique difficulties (e.g., the limited number of available human eggs needed for the cloning process compared to the availability of eggs for animal cloning experiments, etc.).

Within a year of the announcement that mice had been cloned, researchers at Advanced Cell Technology (ACT) in Worcester, Massachusetts, and Kyunghee University in South Korea claimed to have cloned a human embryo from an adult cell for the purposes of embryonic stem cell research.[131] The ACT researchers' embryo was allegedly produced by inserting DNA obtained from a human cheek cell into a cow egg that had been stripped of its genetic material,[132] while the Korean scientists claimed to have produced a clone by combining human DNA with an enucleated human egg cell.[133] ACT's embryo was said to have developed to the thirty-two-cell stage,[134] while the South Koreans' embryo developed to the four-cell stage.[135] Both embryos were then destroyed, making verification of their claims impossible.

In October 2000, researchers at the Australia-based Stem Cell Sciences company and the American firm BioTransplant announced that they had created pig-human embryos by inserting human nuclei into egg cells obtained from pigs. Two of the embryos developed to the thirty-two-cell stage. According to an ABC on-line news report, the scientists' objectives behind cloning the embryos was to develop an alternative to organ donation.[136]

On November 26, 2001, scientists at ACT again announced that they had cloned human embryos—this time using *human* (rather than cow) eggs—for the purpose of developing potentially life-saving medical therapies from the embryos' stem cells (although the embryos failed to develop to the stage at which the coveted stem cells are present).[137] Because none of the embryos developed beyond the six-cell stage—a developmental stage attainable even in the absence of active DNA—some scientists questioned whether cloning had truly occurred.

Additional reports of human embryo cloning have also emerged from Korea. Among these is the Maria Life Engineering Research Institute's March 2002 claim to have produced cloned human embryos by transplanting a human nucleus into a cow egg. The Institute carried out its experiment as a possible means of freeing embryonic stem cell research from its dependence upon human eggs, which can be difficult to obtain.[138] Not to be outdone, a Chinese scientist at Xiangya Medical College also laid claim to cloning and stem cell technology, asserting that she was the first to clone a human embryo, having achieved the feat in 1999.[139] A research team at Shanghai Medical University #2 also claimed to have produced human embryos by fusing human DNA with rabbit eggs as a means of obtaining human embryonic stem cells.[140]

None of the above experiments, though, permitted a cloned human embryo to develop to the point of birth. Although most of the public, as well as the majority of scientists, are outspokenly opposed to this type of cloning, a handful of scientists have declared their intentions to produce live-born human clones. On August 7, 2001, Kentucky-based reproductive physiologist Panos Zavos and Italian infertility doctor Severino Antinori shocked and angered much of the world with their announcement of plans to begin cloning up to two hundred human beings.[141] Also announcing a plan to clone humans was Brigitte

Boissilier, the Director of the organization Clonaid—a group of scientists associated with the Raelians, who believe that the first human beings were the product of a cloning process carried out by an extraterrestrial race.[142] Zavos, who has now parted ways with Antinori, stated in 2002 that Chinese, Russian, and European researchers were making significant progress toward reaching the goal of cloning human beings.[143] That same year, Antinori[144] and Boissilier[145] (with Boissilier speaking on behalf of Clonaid and their Korean affiliate BioFusion) announced that cloned human embryos had already been implanted into women, and Zavos also stated that he was making progress toward his goal of cloning a human being.[146] By early 2003, Boissilier had claimed that multiple human clones had been born, but no scientific evidence to that effect was provided.[147] In April of that year, Zavos announced that he had successfully cloned a human embryo whom he sought to implant in a woman's uterus,[148] only to announce in February 2004 that the woman in whom he had implanted the embryo failed to become pregnant.[149] While the attempt to clone a human embryo has been marked with several presumed failures and false claims, a team of South Korean scientists appears to have succeeded in this pursuit, announcing in February 2004 that they had successfully cloned a human embryo from whom stem cells were derived.[150]

Although human cloning now appears to be technically possible, an overwhelming majority of the public, as well as those within the scientific/medical profession, remain outspoken in their opposition to the birth of cloned babies.[151] Claims that we are moving ever closer to the birth of cloned human beings have prompted a huge outcry from, most notably, the scientific community.[152] The process for cloning animals remains grossly inefficient. In the effort to clone Dolly, it took scientists over 275 attempts before a seemingly healthy sheep was born,[153] and similarly low success rates have been consistently reported for other species that have been cloned.[154] Animal cloning experiments have been plagued by failures over 95 percent of the time, and cloned animals that do survive have most often suffered from grave deformities likely due to the cloning process.[155] Applying this technology to humans would, therefore, almost certainly, prove to be wildly wasteful of human life.

In the event that this process were to prove no longer dangerous, however, public opinion regarding human cloning would likely become more and more positive. Our culture has demonstrated a remarkable ability to reverse its opinions with the passage of time. Many people, for example, were opposed to *in vitro* fertilization (IVF) when it first became available. As time passed and the success rate of the procedure improved, however, most people became accustomed to the technology and began viewing it as acceptable. Human cloning could follow a similar route, with the public eventually coming to view it as "just another reproductive technology." It's important, therefore, that efforts to prevent the development and acceptance of human cloning do not diminish.

38. What is the difference between "reproductive" and "research" (or "therapeutic") cloning?

An overwhelming majority of the public remains opposed to cloning a human being so that a child is actually born, but such wide-scale objection does not greet the prospect of cloning human embryos for the purpose of medical research. Cloning with the former intention has been termed *"reproductive" cloning,* while cloning for the latter purpose has been called "research" (or "therapeutic") cloning. Many scientists and patient advocacy groups strongly support "research" cloning because of its alleged promise to usher in medical breakthroughs that could result in therapies and cures for many of humanity's most debilitating diseases. In seeking to develop such treatments, researchers would obtain stem cells from cloned human embryos to be injected into patients from whom the embryos were cloned. Although some question exists as to whether embryonic stem cells *not* obtained from a patient's clone would in fact be immunologically incompatible and therefore rejected by the patient, cloned embryos have been widely trumpeted as the solution to this potential obstacle. So-called "research" cloning should, however, meet with at least as much disapproval as does "reproductive" cloning, since the former requires the creation and subsequent destruction of cloned human beings (in the embryonic form). Furthermore both types of human cloning are, in reality, "reproductive" human cloning since they result in the creation of a new human life.

39. Why would people want to clone a human being, and is it ever okay to do so?

In answering this question, it's helpful to review the difference between "reproductive" and "research" cloning (see answer to question 38). While both types of cloning produce a cloned human being, the purposes for doing so differ dramatically. "Research" cloning has as its goal not the birth of a child, but the creation of a human being solely for the purpose of harvesting his or her embryonic stem cells—which some people allege could be used to treat or cure the person from whom the clone was created. Proponents of "research" cloning claim that harvesting stem cells from a patient's clone is necessary in order to avoid the possibility of tissue rejection, although "adult" stem cells, which have proven to be surprisingly effective, would also circumvent this problem—which some scientists suggest may even be nonexistent at the stem cell level, regardless of the source from which the cells are derived (see answers to questions 32–34). While the desire to save one's life or the life of a loved one is certainly understandable, we do not believe that "research" cloning offers an ethical means of fulfilling such a desire.

In contrast to "research" cloning, "reproductive" cloning does have as its goal the birth of a child who would be nearly genetically identical to a previously or currently existing individual. Why would someone desire to have such a child? In the aftermath of Dolly, several reasons were offered to justify human cloning in the "reproductive" sense, and some were more fantastic than others.

Initially, many people wondered if human clones could or would be created and brought to term as a source of immunologically compatible vital organs ("spare parts") that would be available for the "original" in the event of an illness or accident. While this could certainly be a possible motive for human cloning, it is highly unlikely that the deliberate killing of clones in order to transplant their hearts or livers into their "originals" would be permitted by any government. What might be greeted with less controversy, however, is the prospect of cloning a human being in order to retrieve immunologically compatible tissue for a patient's bone marrow transplant, kidney transplant, or skin graft—procedures that normally would not harm the clone. Nevertheless, if a clone is not conceived until the need for such treatment arises, the tissue might not be obtainable in time to benefit the patient. Regardless, the decision to have a child—cloned or non-cloned—for the purpose of saving someone else's life should be preceded by serious ethical reflection. Many people would likely consider it a violation of human dignity and worth to "use" clones (or any human being) primarily for other people's benefit.

The most frequently offered justification for human cloning is the desire to help childless couples have their own genetically related children. Despite all that can be done today to combat infertility, some couples still cannot have children of their own, either because of defective sperm or eggs or both. Cloning proponents point out that a way for these couples to have genetically related children is to have either the husband or wife (or perhaps both) cloned. This option might seem particularly attractive in that it would eliminate the need for donor sperm and/or eggs (which some couples may find objectionable) and would enable infertile couples to have a child who is genetically related (in fact, nearly identical) to one of the spouses. Imagine, though, the psycho-sociological dimension—intriguing at best, destructive at worst—in homes where a child was nearly genetically identical to his or her father or mother. Moreover, the cloned baby would actually be (genetically speaking) the child of the cloned person's *parents,* rather than the genetic offspring of the cloned person and his or her spouse. Cloning would not, therefore, provide an infertile couple with *their* own child—the clone would in fact be a twin brother or sister of the cloned spouse and would bear no genetic relationship at all to the other spouse.

Cloning might also provide prospective parents with a means of preventing their children from inheriting a particular genetic illness that one or both spouses was at risk of passing on. If, for example, a husband was known to carry a dominant disease-causing gene, his wife could be cloned. Or if both spouses had a dominant disease-causing gene, they could arrange for the wife to carry a clone of someone else (such as a favorite relative). These are not the only ways, however, that such a couple could raise or even give birth to a child free from genetic disease. Many healthy embryos, for example, are frozen and are waiting to be adopted and carried to term (see question 34).[156]

Couples who had tragically lost a child might view cloning as a way to "replace" their beloved son or daughter. But given what we know about how environment shapes individuals, the cloned child would likely differ significantly in many ways from the child who had died. Nevertheless, unrealistic expectations would likely be placed on the clone to exhibit certain behaviors and personality traits. Regardless of the outcome, attempting to replace a lost child via cloning would be an inappropriate way to deal with grief.

Finally, couples desiring an especially athletic son or a musically talented daughter might seek to clone celebrities such as Michael Jordan or Celine Dion. Although clones of such personalities would likely have very different abilities and traits, some people would very likely wish to clone as a result of this motivation. Many parents, of course, already have children primarily for self-serving reasons without resorting to cloning. Such instances, however, are normally impossible to document and thereby prevent—although they are no more ethical than in the context of cloning.

While some of the above reasons for engaging in "reproductive" cloning might appear less objectionable than others—and may by some people even be regarded as wholly justifiable—it is our belief that human cloning is *intrinsically wrong* for several reasons. First, children are not products to be designed; rather, they are creations of God and their own interests must be taken seriously.[157]

Furthermore, children's interests are typically best served by having a genetic and a relational connection to their parents. Human cloning denies the child a complete set of genetic parents and thus disadvantages him or her—even if the pursuit of other interests being served by the cloning does not otherwise harm the child. A couple's joint interests are also best served by sharing a genetically and relationally similar relationship with their children that is as complete as possible. Thus, the parents themselves would also likely be disadvantaged if one has a major genetic connection to the child and the other does not. Such an imbalance could easily undermine the stability of the parents' relationship with each other and with the child. While genetics is not all there is to the parent-child relationship, it is a natural basis for it.[158]

Among other objections to human cloning is that perfecting this process will almost certainly require the sacrifice of many human embryos in the endeavor to bypass disease, alleviate parental grief, or provide a couple with their "dream child." Should cloning one day become as safe as the natural means of reproduction, it is still highly likely that many human embryos would have lost their lives (without their consent) in the name of scientific progress. Furthermore, cloning would involve the imposition of a genetic code—chosen by others—on a child, a prospect that should give us great pause given the permanent implications of a child's genetic traits. Aside from a cloned human being's genetic make up, clones—who were created with the hope that they would mirror the life of the individual from whom they were cloned—might feel less freedom to develop their own interests than would non-cloned people. Such impositions would constitute a basic violation of human dignity. Finally, human cloning represents a fundamental break with God's design for human procreation—far more so than does IVF or any other reproductive technology (which still seek to join sperm and egg in the creation of a new life). Such a departure is worthy of serious consideration by would-be cloners.

In summary, the creation of a cloned human being is ethically unacceptable, whether for "research" or "reproductive" purposes.[159]

40. Would clones be fully human?

Most people do not believe that there is any physical reason to assume that clones would be less than fully human, since identical twins are (without question) fully human. Clones created from human DNA and a human egg cell would, indeed, be biologically human, but clones and identical twins originate differently, and the full implications of that difference are not known.

A related dimension concerns the question of whether human clones would have souls. The question of "ensoulment"—i.e., how a human being comes to possess a soul—has long intrigued theologians. The theory of *traducianism* holds that a human being's soul (as well as his or her physical nature) is transmitted directly by his or her parents. The *creationist* theory, on the other hand, maintains that, while a person's parents bestow a physical nature on their child, the soul "is specially created by God for each individual and united with the body."[160] The traducianist theory might cast some doubt on the claim that human clones would have souls, depending upon how precise and confident the theory is about what biological material from the parents the soul attaches to. The problem would be that the material used to create a clone (a body cell containing forty-six chromosomes obtained from a single individual plus an enucleated egg cell) is not the same material used in God's original design for the creation of human life (a man's sperm containing twenty-three chromosomes and a woman's egg containing twenty-three chromosomes). The creationist theory would likely pose fewer problems for the assertion that human clones would have souls, since God bestows the soul apart from the creation of the body.

Some may assume that human clones would have souls, since identical twins have souls. That is, because an identical twin also does not result directly from the fusion of a sperm and egg—but unquestionably has a soul—human clones would likewise have souls. While it remains true that twins do originate as part of God's design, clones are created in a manner wholly distinct from God's procreative plan. Furthermore, some people believe that two souls are present from the time of conception in an embryo that later divides.

Because the Bible is not explicit about how our souls become a part of who we are, it is understandable that the question should arise—would clones have souls? While it's not our position that a clone would lack a soul, we have offered the above material to illustrate that the answer to this question may not be as simple as many people believe.

We can't say with certainty that the arrival of the soul or spirit is not tied somehow to the method of procreation. However, the Bible is consistent in its view that the material and non-material aspects of our being are a unified whole. To have a spiritual existence without a body (even after death), for example, is described as being "naked" (2 Cor. 5:3)—which is partly why the resurrection of the body is described as important Christian doctrine (see 1 Cor. 15; 2 Cor. 5). Since the Bible does emphasize a holistic picture of people—affirming that both the spiritual and material dimensions are necessary components of human beings—it would not seem unreasonable to assume that when a live, genetically human individual is present, a soul is also present.

Regardless of the correct theory of ensoulment/humanness, cloning would be a radically new way of bringing human beings into the world. While God is ultimately the determiner of whether cloned human beings would have souls and would be fully human, we must protect as human any beings that may well be so. We must take steps to ensure that such individuals—once created—are treated in every respect as if they are fully human, rather than being subjected to abuse or regarded as inferior to non-cloned human beings.

Conclusion

Many people today fail to see the obvious fingerprint of God in the design of human life—even with increasing knowledge of the amazing intricacies of the human genome. Yet God is allowing science to uncover the workings of His mind and creative power more than ever before. While such new understanding may well pave the way for advances in genetics, transgenics, xenotransplantation, stem cell research, and cloning, the magnificence of what is learned should point people ultimately to Him—with the result that these technologies will be employed in ways that are God-honoring. For Christians, such knowledge is, indeed, another indicator of God's majesty and glory and should stir up in each of us a greater desire and commitment to worship and serve Him (cf. 1 Chron. 29:10–16).

Our expanding glimpse of the human genome, coupled with a myriad of other scientific advances, makes the words of the psalmist even more revealing:

> For you created my inmost being;
> you knit me together in my mother's womb.
> I praise you because I am fearfully and wonder-
>   fully made;
> your works are wonderful,
> I know that full well.
> My frame was not hidden from you
> when I was made in the secret place.
> When I was woven together in the depths of the
>   earth,
> your eyes saw my unformed body.

All the days ordained for me
were written in your book
before one of them came to be.
How precious to me are your thoughts, O God!
How vast is the sum of them!
Were I to count them,
they would outnumber the grains of sand.
When I awake
I am still with you.

—Psalm 139:13–18

It is our hope that this booklet will equip people to increase their knowledge of genetics, as well as the various other biotechnologies discussed in these pages, so that the decisions and consequences surrounding these challenging issues will ultimately honor God, serve people, and care for His creation.

Recommended Resources

The Center for Bioethics and Human Dignity Resources:

Kilner, John F., and C. Ben Mitchell. *Does God Need Our Help? Cloning, Assisted Suicide, and Other Challenges in Bioethics.* Wheaton, Ill.: Tyndale, 2003.

Kilner, John F., et al., eds. *Cutting-Edge Bioethics: A Christian Exploration of Technologies and Trends.* Grand Rapids: Eerdmans, 2002.

Kilner, John F., et al., eds. *The Reproduction Revolution: A Christian Appraisal of Sexuality, Reproductive Technologies, and the Family.* Grand Rapids: Eerdmans, 2000.

Kilner, John F., et al., eds. *Genetic Ethics: Do the Ends Justify the Genes?* Grand Rapids: Eerdmans; and United Kingdom: Paternoster, 1997.

Other Resources:

Demy, Timothy J., and Gary P. Stewart, eds. *Genetic Engineering: A Christian Response: Crucial Considerations in Shaping Life.* Grand Rapids: Kregel, 1999.

Prentice, David A., and Michael A. Palladino. *Stem Cells and Cloning.* San Francisco: Benjamin Cummings, 2002.

Ramsey, Paul. *Fabricated Man: The Ethics of Genetic Control.* New Haven and London: Yale University Press, 1970.

Song, Robert. *Human Genetics: Fabricating the Future.* Cleveland: Pilgrim Press, 2002.

Endnotes

1. "International Consortium Completes Human Genome Project," available at http://www.nih.gov/news/pr/apr2003/nhgri-14.htm (accessed 9 March 2004).
2. CNN.com, "Scientists Sequence First Human Chromosome," available at http://www.cnn.com/1999/HEALTH/12/01/chromosome.22/ (accessed 20 October 2002).
3. Michael Shamblott et al., "Derivation of Pluripotent Stem Cells from Cultured Human Primordial Germ Cells," *Proceedings of the National Academy of Sciences of the United States of America* 95 (November 1998): 13726–31; and James Thomson et al., "Embryonic Stem Cell Lines Derived from Human Blastocysts," *Science* 282 (6 November 1998): 1145–47.
4. See www.stemcellresearch.org.
5. Gretchen Vogel, "Scientists Take Step Toward Therapeutic Cloning," *Science* 303 (13 February 2004): 937–39.
6. See CNN.com, "Raelian Leader Says Cloning First Step to Immortality," 28 December 2002, available at http://www.cnn.com/2002/HEALTH/12/27/human.cloning/ (accessed 6 November 2003); and CNSNews.com, "Claims of Another Clone Birth Prompt Skepticism, Concern," 12 February 2004, available at www.CNSNews.com (accessed 9 March 2004).
7. Francis S. Collins, "Testimony Before the Appropriations Subcommittee on Labor, Health and Human Services and Education, United States Senate,"

11 July 2001, available at www.genome.gov (accessed 20 October 2002).

8. Carl Weiland, "Man's Achievements vs. Amazing 'Living Computer' Technology," *Creation ex Nihilo* 21, no. 1 (December 1998–February 1999): 10–11. Originally cited in Werner Gitt, *The Wonder Man* (Germany: Christliche Literatur-Verbreitung; December 1999).

9. Weiland, "Man's Achievements vs. Amazing 'Living Computer' Technology."

10. Ibid., 10–11. Originally cited in Jerome LeJeune, "Anthropotes, Rivista di studi sulla persona e la famiglia. Citta Nuova Editrice," 1989.

11. Francis Collins, "Human Genetics," in *Cutting-Edge Bioethics: A Christian Exploration of Technologies and Trends,* ed. John F. Kilner, et al. (Grand Rapids: Eerdmans, 2002).

12. See www.genome.gov (accessed 20 October 2002).

13. Ibid.

14. The White House Office of the Press Secretary, "President Clinton Announces the Completion of the First Survey of the Human Genome," 25 June 2000 press release, available at http://www.ornl.gov/hgmis/project/clinton1.html (accessed 20 October 2002).

15. *Science* (15 February 2001): Special Issue.

16. "Initial Sequencing and Analysis of the Human Genome." *Nature* 409 (15 February 2001): 860–921.

17. "International Consortium Completes Human Genome Project."

18. Francis Collins, "Contemplating the End of the Beginning," *Genome Research* 11 (May 2001): 641–43.

19. Ray Bohlin, "Genetic Intervention: The Ethical Challenges Ahead," *Dignity* newsletter (Deerfield, Ill.: Center for Bioethics and Human Dignity, spring 2002), available at www.cbhd.org (accessed 20 October 2002).

20. Leroy B. Walters, "Behavioral and Germ-line Genetic Research," in *Genetic Ethics: Do the Ends Justify the Genes?* ed. John F. Kilner, et al. (Grand Rapids: Eerdmans, 1997), 105–8.

21. J. Daryl Charles, "Blame It on the Beta-Boosters: Genetics, Self-determination, and Moral Accountability," in *Genetic Engineering: A Christian Response,* ed. Timothy J. Demy and Gary P. Stewart (Grand Rapids: Kregel, 1999), 248.

22. Dorothy Nelkin and M. Susan Lindee, *The DNA Mystique: The Gene as a Cultural Icon* (New York: W. H. Freeman and Co., 1995), 39.

23. Collins, "Human Genetics," 4, 16.

24. Nelkin and Lindee, *The DNA Mystique: The Gene as a Cultural Icon.*

25. Dean H. Hamer, S. Hu, V. L. Magnuson, N. Hu, and A. M. Pattatuci, "A Linkage Between DNA Markers on the X Chromosome and Male Sexual Orientation," *Science* 261 (1993): 321–25; and Dean H. Hamer and Peter Copeland, *The Science of Desire: The Search for the Gay Gene and the Biology of Behavior* (New York: Simon and Schuster, 1994), 272.

26. M. Bailey and R. C. Pillard, "A Genetic Study of Male Sexual Orientation," *Archives of General Psychiatry* 48 (1991): 1089–96; and S. LeVay, "A Difference in Hypothalamic Structure Between Heterosexual and Homosexual Men," *Science* 253 (1991): 1034–376.

27. See Jeffrey Satinover. *Homosexuality and the Politics of Truth*. Grand Rapids: Baker Books, 1996; and http://www.narth.com/docs/istheregene.html (accessed 15 March 2004).

28. N. Risch, et al. "Male Sexual Orientation and Genetic Evidence." (Letter) *Science* 262 (1993): 2063–65.

29. G. Rice, C. Anderson, N. Risch, and G. Fibers, "Male Homosexuality: Absence of Linkage to Microsatellite Markers at Xq28," *Science* 284 (1999): 665–67; and

Ingrid Wickelgren, "Discovery of 'Gay Gene' Questioned," *Science* 284 (1999): 571.

30. *Gattaca,* videocassette, directed by Andrew Niccol (Los Angeles: Jersey Films, 1997).

31. Paul Berg, et al., *Nature* 250 (1974): 175; and idem, *Science* 185 (1974): 303.

32. "Plant Biotechnology: Food and Feed," special section in *Science* 285 (16 July 1999): 367–89.

33. See Genetically Engineered Food Alert's "Press Room," available at http://www.gefoodalert.org/takeaction/html/pressroom.htm (accessed 20 October 2002).

34. Pew Initiative on Food and Biotechnology, "Pew Initiative on Food and Biotechnology Finds Public Opinion About Genetically Modified Foods 'Up for Grabs,'" 26 March 2001, available at http://pewagbiotech.org/newsroom/releases/ (accessed 20 October 2002).

35. Rutgers News, "To the Point: Rutgers' Food Policy Institute Report Shows that Americans [sic] Confused and Undecided About Biotechnology," 15 November 2001 press release, available at http://www.cook.rutgers.edu/www/news/pressreleases (accessed 20 October 2002).

36. Michael W. Fox, *Beyond Evolution: The Genetically Altered Future of Plants, Animals, the Earth . . . and Humans* (New York: Lyons Press, 1999), 117–18.

37. Ibid., 118–19.

38. See www.alzheimers.org/genefact.html (accessed on 20 October 2002).

39. James Shreeve, "Secrets of the Gene," *National Geographic* 196 (October 1999): 62.

40. Ibid., 57.

41. Elizabeth Thomson, "Genetic Counseling," in *Genetic Ethics: Do the Ends Justify the Genes?*

42. Ibid.

43. Ibid.

44. Ibid.

45. Ibid.

46. C. Christopher Hook, "Genetic Testing and Confidentiality," in *Genetic Ethics: Do the Ends Justify the Genes?*

47. Ibid., 127–28.

48. Ibid., 132.

49. Shreeve, "Secrets of the Gene," 64–65.

50. C. Ben Mitchell, "Genetic Engineering: Bane or Blessing," in *Genetic Engineering: A Christian Response*, 37.

51. "Gene Therapy," available at www.ornl.gov/hgmis/medicine/genetherapy.html (accessed 20 October 2002).

52. "National Cancer Institute Cancer Facts," 7 June 2000, available at http://cis.nci.nih.gov/fact/7_18.htm (accessed 20 October 2002).

53. Leon Jaroff, "Success Stories," *Time,* 11 January 1999, 72–73.

54. Sally Lehrman, "Virus Treatment Questioned After Gene Therapy Death," *Nature* 401 (7 October 1999): 517–18.

55. Eliot Marshall, "FDA Halts All Gene Therapy Trials at Penn," *Science* 287 (28 January 2000): 565–67.

56. Meredith Wadman, "NIH Under Fire over Gene-Therapy Trials," *Nature* 403 (20 January 1999): 237.

57. Andrew Pollack, "Gene Therapy Trials Halted," *New York Times,* 15 January 2003, available at http://www.nytimes.com/2003/01/15/health/15GENE.html (accessed 17 January 2003).

58. See www.ornl.gov.

59. Ricki Lewis, "Preimplantation Genetic Diagnosis: The Next Big Thing?" *The Scientist* (13 November 2000).

60. BBC News, "Designer Baby Born to UK Couple," 19 June 2003, available at http://news.bbc.co.uk (accessed 16 March 2004).

61. Kirsty Horsey, "Hashmis Can Go Ahead with Embryo Tissue-Typing," 15 April 2003, http://www.ivf.net (accessed 16 March 2004).

62. Some helpful web sites are those of the Directory of Online Genetic Support Groups, www.mostgene.org/support/ and of the Genetic Alliance, www.geneticalliance.org (accessed 20 October 2002).

63. Carol M. Ostrom, "New Technique Lets Parents Pick Baby's Gender," *Seattle Times,* 16 October 2002, available at http://seattletimes.nwsource.com/html/localnews/134556028_babies16m.html (accessed 20 October 2002).

64. See Mary A. Kassian, "Biology Equals Destiny," in *The Feminist Gospel: The Movement to Unite Feminism with the Church* (Wheaton, Ill.: Crossway Books, 1992), 43–50; and idem, "The First Sex," in *The Feminist Gospel,* 99–107.

65. Lee Silver, *Remaking Eden: How Genetic Engineering and Cloning Will Transform the American Family* (New York: Avon Books, 1998), 188.

66. Robert Wright, "Who Gets the Good Genes?" *Time,* 11 January 1999, 67.

67. John S. Feinberg, "A Theological Basis for Genetic Intervention," in *Genetic Ethics: Do the Ends Justify the Genes?*

68. Michael D. Lemonick, "Smart Genes?" *Time,* 13 September 1999, 54–58.

69. Nancy Gibbs, "If We Have It Do We Use It?" *Time,* 13 September 1999, 59–60.

70. Edward J. Larson, "Confronting Scientific Authority With Religious Values: Eugenics in American History," in *Genetic Engineering.*

71. Ibid., 112.

72. Oliver Wendell Holmes, quoted by Paul Gray, "Cursed by Eugenics," *Time,* 11 January 1999, 85.

73. Arthur J. Dyck, "Eugenics in Historical and Ethical

Perspective," in *Genetic Ethics: Do the Ends Justify the Genes?*

74. Ibid.

75. Edward O. Wilson, *Consilience: The Unity of Knowledge* (New York: Alfred Knopf, 1998), 146.

76. Silver, *Remaking Eden: How Genetic Engineering and Cloning Will Transform the American Family,* 1–7.

77. Allen D. Verhey, "Playing God," in *Genetic Ethics: Do the Ends Justify the Genes?*

78. U.S. Congress, Office of Technology Assessment, *New Developments in Biotechnology: Patenting Life—Special Report*, OTA-BA-370 (Washington, D.C.: U.S. Government Printing Office, 1989), 7.

79. *Diamond v. Chakrabarty*, 447 U.S. 303, 65 L. Ed. 2d 144, 100 S. Ct. 2204.

80. *Ex parte Allen,* 2 USPQ2d 1425.

81. *Moore v. Regents of the University of California,* 793 P.2d 479 (Cal. 1990).

82. Paige C. Cunningham, "The Right to Patent a Human Being: Fact, Fiction, or Future Possibility?" the Center for Bioethics and Human Dignity, 2002, available at www.cbhd.org (accessed 20 October 2002).

83. Martin Ensirink, "Patent Office May Raise the Bar on Gene Claims," *Science* 287 (18 February 2000): 1196–97.

84. C. Ben Mitchell, "Patenting Life: An Evangelical Response," in *Genetic Engineering,* 99–100.

85. Nancy L. Jones and Linda Bevington, "Transgenics," in *Cutting-Edge Bioethics.*

86. Fox, *Beyond Evolution,* 117–18.

87. Shreeve, "Secrets of the Gene," 44–45.

88. Applied Genetics News, "Some Pigs Are More Equal than Others," April 2000.

89. Rachel Nowak, "Almost Human," *New Scientist,* 13 February 1999.

90. C. Sibley and J. Ahlquist, *Journal of Molecular Evolution* 26 (1987): 99–121; and M. Sarich et al., "DNA Hybridization as a Guide to Phylogenics: A Critical Analysis," *Cladistics* 5 (1989): 3–32. *Text Note:* While this percentage may seem large, remember that we possess 3 billion base pairs of DNA in our chromosomes; therefore, many millions of them are *not* shared by apes. Rather than being negligible, this is actually a huge quantifiable difference. Furthermore, this percentage is not based upon a complete sequencing of chimpanzee DNA, but upon an inconclusive technique known as DNA hybridization, and is therefore somewhat arbitrary.

91. Rick Weiss, "Pigs Cloned in Transplant Program," *Boston Globe,* 15 March 2000, A3.

92. See www.ppl-therapeutics.com/news/news_1.html (accessed 20 October 2002); and Liangxue Lai, "Production of alpha-1,3-Galactosyltransferase Knockout Pigs by Nuclear Cloning," *Science* 295 (3 January 2002): 1089–92.

93. "Researchers Present Data on First Cloned, Double Knock-out Miniature Swine," 13 January 2003, available at http://www.immergebt.com/press_room/2003_01_13.php (accessed 18 November 2003).

94. "Infigen Announces the Birth of Genetically Modified Miniature Swine for Potential Use as Organ Donors for Humans," 27 February 2003, available at http://www.infigen.com/news/news_20030227.html (accessed 18 November 2003).

95. Marialuisa Lavitrano et al., "Efficient Production by Sperm-Mediated Gene Transfer of Human Decay Accelerating Factor (hDAF) Transgenic Pigs for Xenotransplantation," *Proceedings of the National Academy of Sciences of the United States of America,* 22 October 2002, available at www.pnas.org/papbyrecent.shtml (accessed 20 October 2002).

96. The key molecule for PERV infection has now been

identified—an important advance in making xeno-transplantation a reality. "Immerge BioTherapeutics Announces Identification of Key Molecule Responsible for Porcine Endogenous Retrovirus (PERV) Infection," 26 May 2003, available at http://www.immergebt.com/press_room/2003_05_26.php (accessed 18 November 2003).

97. David Cook, "Xenotransplantation," in *Cutting-Edge Bioethics: A Christian Exploration of Technologies and Trends*, 2002.

98. Ernst Haeckel, *The Wonders of Life,* trans. J. McCabe (London: Watts, 1905), 111; cited by Bert Thompson, "Cracking the Code: The Human Genome Project in Perspective," *Apologetics Press* 20, no. 8 (2000): 57–58.

99. William P. Cheshire Jr., "Human Embryo Research After the Genome," The Center for Bioethics and Human Dignity, 2002, available at www.cbhd.org (accessed 19 February 2003).

100. William P. Cheshire Jr., "The Ethics of Human Parthenogenesis," Christian Medical Association "White Paper," 2002.

101. Sec. 510(b) of P.L. 107–116 (Labor/HHS/Education Appropriations Act for FY 2002).

102. "The Dickey Amendment," current version, H.R. 3061, Public Law 107–16, Section 510, Departments of Labor, Health and Human Services, and Education, and Related Agencies Appropriations Act, 2002.

103. National Bioethics Advisory Commission, *Cloning Human Beings* (Rockville, Md.: June 1997), 1:A-2; and National Academy of Sciences, *Stem Cells: Scientific Progress and Future Research Directions* (National Institutes of Health, June 2001), F-3.

104. National Academy of Sciences, *Stem Cells: Scientific Progress and Future Research Directions,* E-5.

105. Cheshire, "Human Embryo Research After the Genome."

106. John F. Kilner and C. Ben Mitchell, *Does God Need Our Help? Cloning, Assisted Suicide, & Other Challenges in Bioethics* (Wheaton, Ill.: Tyndale, 2003).

107. Michael Shamblott, et al., "Derivation of Pluripotent Stem Cells from Cultured Human Primordial Germ Cells," 13726–31; and James Thomson et al., "Embryonic Stem Cell Lines Derived from Human Blastocysts," 1145–47.

108. Michael Shamblott, et al., "Derivation of Pluripotent Stem Cells from Cultured Human Primordial Germ Cells," 13726–31.

109. James Thomson, et al., "Embryonic Stem Cell Lines Derived from Human Blastocysts," 1145–47.

110. S. Wakitani, et al., "Embryonic Stem Cells Injected into the Mouse Knee Joint Form Teratomas and Subsequently Destroy the Joint," *Rheumatology* (January 2003) 42:162–65.

111. C. R. R. Bjornson, et al., "Turning Brain into Blood: A Hematopoietic Fate Adopted by Adult Neural Stem Cells in Vivo," *Science* 283 (22 January 1999): 534.

112. Yuehua Jing, et al., "Pluripotency of Mesenchymal Stem Cells Derived from Adult Marrow," *Nature* 418 (4 July 2002): 41–49.

113. United States Conference of Catholic Bishops, "Current Clinical Use of *Adult* Stem Cells to Help Human Patients," available at www.nccbuscc.org/prolife/issues/bioethic/adult701.htm (accessed 20 October 2002).

114. M. Hakelien, et al., "Reprogramming Fibroblasts to Express T-cell Functions Using Cell Extracts," *Nature Biotechnology* 20 (May 2002): 460–66.

115. Robert W. Evans, "The Moral Status of Embryos," in *The Reproduction Revolution: A Christian Appraisal of Sexuality, Reproductive Technologies, and the Family,* ed. John F. Kilner et al. (Grand Rapids: Eerdmans, 2000).

116. The Center for Bioethics and Human Dignity, "On Human Embryos and Stem Cell Research: An Appeal for Legally and Ethically Responsible Science and Public Policy," 1 July 1999, available at www.cbhd.org (accessed 20 October 2002).

117. John and Lucinda Borden, "Testimony Before the U.S. House of Representatives Committee on Government Reform Subcommittee on Criminal Justice, Drug Policy, and Human Resources Hearing on Embryonic Stem Cell Research," 17 July 2001, available at http://www.stemcellresearch.org/testimonies/borden.htm (accessed 20 October 2002).

118. Linda K. Bevington, "Federally Funding Embryonic Stem Cell Research: Bush and Beyond," *Dignity* newsletter, fall 2001 (Bannockburn, Ill.: Center for Bioethics and Human Dignity), available at www.cbhd.org (accessed 20 October 2002).

119. Ibid.

120. United States Conference of Catholic Bishops, "Fact Sheet: Embryonic Stem Cell Research and Vaccines Using Fetal Tissue," available at www.nccbuscc.org/prolife/issues/bioethic/vaccfac2.htm (accessed 20 October 2002).

121. The Center for Bioethics and Human Dignity, "On Human Embryos and Stem Cell Research."

122. John Gurdon, "The Developmental Capacity of Nuclei Taken from Intestinal Epithelium Cells of Feeding Tadpoles," *Journal of Embryology and Experimental Morphology* 10 (1962): 622–40.

123. John Gurdon, et al., "The Developmental Capacity of Nuclei Transplanted from Keratinized Skin Cells of Adult Frogs," *Journal of Embryology and Experimental Morphology* 34 (1975): 93–112.

124. Ian Wilmut, et al., "Viable Offspring Derived from Fetal and Adult Mammalian Cells," *Nature* 385 (1997): 810–13.

125. Ibid.

126. Elizabeth Pennisi, "Transgenic Lambs From Cloning Lab," *Science* 277 (1997): 631.

127. Yoshimi Kuroiwa et al., "Cloned Transchromosomic Calves Producing Human Immunoglobulin," *Nature Biotechnology* (published online 12 August 2002).

128. Genetic Savings and Clone, available at www.savingsandclone.com (accessed 20 October 2002).

129. Ibid.

130. Teruhiko Wakayama et al., "Mice Cloned from Adult Cell Nuclei," *Nature* 394 (23 July 1998): 369–73.

131. Advanced Cell Technology, "Advanced Cell Technology Announces Use of Nuclear Transfer Technology for Successful Generation of Human Embryonic Stem Cells," *Press Release,* 12 November 1998, available at http://www.advancedcell.com/pr_11-12-1998.html (accessed 26 February 2003); and BBC News, "Scientists Make Human Clone Claim," 16 December 1998, available at http://news.bbc.co.uk/1/hi/sci/tech/236089.stm (accessed 26 February 2003).

132. Advanced Cell Technology, "Advanced Cell Technology Announces Use of Nuclear Transfer Technology."

133. BBC News, "Scientists Make Human Clone Claim."

134. "Human Cloning from Human Cell and Cows [sic] Egg: World's Best Kept Secret in Cloning Research," available at http://www.globalchange.com/humancow.htm (accessed 26 February 2003).

135. BBC News, "Scientists Make Human Clone Claim."

136. ABC Science Online, "Human-Pig Embryo Accusation Provokes Debate," 9 October 2000, available at www.abc.net.au/science/news/stories/s196491.htm. (accessed 20 October 2002).

137. Jose P. Cibelli et al., "The First Cloned Human Embryo," *Scientific American* (January 2002).

138. Yang Sung-jin, "Korean Scientists Clone Human Embryo Using Cow Eggs," *The Korea Herald,* 9

March 2002, available at www.koreaherald.com (accessed 20 October 2002).

139. Karby Leggett, "China Makes Embryo Cloning Its Baby," *Financial Review,* 8 March 2002, available at www.afr.com/asia/2002/03/08 (accessed 20 October 2002).

140. LifeSite Daily News, "China Boasts of Human Cloning and Human-rabbit Embryos," 6 March 2002, available at www.lifesite.net/ldn/2002/mar/02030603.html (accessed 20 October 2002).

141. Graham Jones, "First Human Clone Bid Planned," 7 August 2001, available at http://www.cnn.com/2001/WORLD/europe/08/06/clone.doctor/ (accessed 26 February 2003).

142. CNN.com, "Scientists Blast Human Cloning Plans," 8 August 2001, available at http://www.cnn.com/2001/WORLD/europe/08/07/cloning/ (accessed 26 February 2003).

143. NewScientist.com News Service, "2002 to Be 'Year of the Clone,' Claims Scientist," 16 May 2002.

144. Americans to Ban Cloning, "Reuters: Antinori Claims Three Clonal Pregnancies, One in 10th Week," 8 May 2002 [Cloning Fact], available at http://www.cloninginformation.org/info/cloningfact/fact-02-05-08.htm (accessed 26 February 2003).

145. ABCNews Online, "Raelians Launch First Attempt at Human Cloning," 15 April 2002, available at http://www.abc.net.au/news/scitech/2002/04/item20020412110456_1.htm (accessed 26 February 2003).

146. David Brown, "Human Clone's Birth Predicted: Delivery Outside U.S. May Come by 2003, Researcher Says," *The Washington Post,* 16 May 2002, available at http://www.washingtonpost.com/ac2/wp-dyn?pagename=article&node=&contentId=A24083-2002May15&notFound=true (accessed 26 February 2003).

147. CNN.com, "Clone Experts Scoff at Baby Claims," 6 January 2003, available at http://www.cnn.com/2003/HEALTH/01/05/human.cloning/ (accessed 26 February 2003).

148. Nell Boyce, "A Clone at Last?" *U.S. News & World Report,* 21 April 2003.

149. BBC News, "Human Cloning Attempt Has Failed," 4 February 2004, available at http://news.bbc.co.uk (accessed 15 March 2004).

150. Gretchen Vogel, "Scientists Take Step Toward Therapeutic Cloning," *Science* 303 (13 February 2004): 937–39.

151. The President's Council on Bioethics, "Human Cloning and Human Dignity: An Ethical Inquiry," Washington, D.C.: July 2002, available at http://www.bioethics.gov/reports/cloningreport/index.html (accessed 26 February 2003); and Gallup poll, 16 May 2002.

152. Rudolf Jaenisch et al., "Don't Clone Humans!" *Science* 291 (2001): 2552.

153. Wilmut et al., "Viable Offspring Derived from Fetal and Adult Mammalian Cells," 810–13.

154. Jaenisch et al., "Don't Clone Humans!" 2552.

155. Ibid.

156. Snowflakes Embryo Adoption Program, available at www.snowflakes.org (accessed 20 February 2003).

157. For further discussion of this matter, see Gary P. Stewart et al., *Basic Questions on Sexuality and Reproductive Technology,* BioBasics Series (Grand Rapids: Kregel, 1998), questions 5 and 6.

158. We recognize that adoption is often a wonderful response to a tragic situation; however, the motivations behind and circumstances unique to cloning do not make it a viable means of welcoming a child into a family.

159. For a more in-depth discussion and evaluation of human cloning, see John F. Kilner, "Human Cloning,"

in *The Reproduction Revolution;* and William P. Cheshire Jr. et al. "Stem Cell Research: Why Medicine Should Reject Human Cloning," *Mayo Clinic Proceedings* 78 (August 2003): 1010–18.

160. Millard Erickson, *Christian Theology* (Grand Rapids: Baker, 1983).